EXCITON AND DOMAIN LUMINESCENCE OF SEMICONDUCTORS

EKSITONNAYA I DOMENNAYA LYUMINESTSENTSIYA POLUPROVODNIKOV

ЭКСИТОННАЯ И ДОМЕННАЯ ЛЮМИНЕСЦЕНЦИЯ ПОЛУПРОВОДНИКОВ

The Lebedev Physics Institute Series

Editors: Academicians D. V. Skobel'tsyn and N. G. Basov

P. N. Lebedev Physics Institute, Academy of Sciences of the USSR

Recent Volumes in this Series

Proceedings (Trudy) of the P. N. Lebedev Physics Institute

Volume 97

Exciton and Domain Luminescence of Semiconductors

Edited by
N. G. Basov

P. N. Lebedev Physics Institute
Academy of Sciences of the USSR
Moscow, USSR

Translated from Russian by
R. C. Hutchison

CONSULTANTS BUREAU
NEW YORK AND LONDON

Library of Congress Cataloging in Publication Data

Main entry under title:

Exciton and domain luminescence of semiconductors.

(Proceedings (Trudy) of the P. N. Lebedev Physics Institute; v. 97).
Translation of Eksitonnaiâ i domennaiâ liûminestŝentŝiiâ poluprovodnikov.
Includes index.
1. Semiconductors—Addresses, essays, lectures. 2. Luminescence—Addresses, essays, lectures. 3. Exciton theory—Addresses, essays, lectures. 4. Domain structure—Addresses, essays, lectures. I. Basov, Nikolaĭ Gennadievich, 1922-   II. Series: Akademiiâ nauk SSSR. Fizicheskiĭ institut. Proceedings; v. 97.
QC1.A4114 vol. 97 [QC611.2]   530′.08s [537.6′]   79-14567

ISBN-13:978-1-4615-8575-6   e-ISBN-13:978-1-4615-8573-2
DOI: 10.1007/978-1-4615-8573-2

The original Russian text was published by Nauka Press in Moscow in 1977 for the Academy of Sciences of the USSR as Volume 97 of the Proceedings of the P. N. Lebedev Physics Institute. This translation is published under an agreement with the Copyright Agency of the USSR (VAAP).

PREFACE

This volume, which comprises a collection of papers by leading Soviet researchers, is devoted to topics in the luminescence of semiconductors. An experimental check is made on a series of predictions of the theory of ionization domains. A new low-voltage luminescence of zinc sulfide is described and investigated and is found to be due to a high-frequency electrical instability. A detailed study of the electrical properties of the instability and of the characteristics of the emission testifies to the pre-breakdown character of the electroluminescence and to the acousto-electrical nature of the instability. The luminescence excitation spectra of AlN crystals excited in the region of the fundamental absorption contain lines belonging to excitons and their phonon replicas. The symmetry of the electronic and vibrational transitions corresponding to parts of these lines is interpreted. The results of a study of the scattering of light by electron—hole drops in germanium are cited. The results are discussed on the basis of a theory of exciton condensation in which allowance is made for the diffusion of excitons toward the surface of the drops and for the surface tension of the electron—hole liquid. This volume will be of interest to a wide range of scientific workers, particularly those engaged in the study of luminescence and physics of semiconductors.

CONTENTS

CONTENTS

THE THEORY OF IONIZATION DOMAINS.
GENERALIZATION AND EXPERIMENTAL CHECK

M. V. Fok, E. V. Devyatykh, and E. Yu. L'vova

The behavior of electrons of energy greater than 1 eV is investigated not only in fields of the order of 10^6 V/cm but also in weaker fields, up to 10^3 V/cm. New confirmation of the theory of ionization domains is obtained: the domain velocity increases as the domain approaches the anode, and there is an accompanying increase in the current through the domain. The processes in the near-cathode space-charge layer that lead at large fields to the appearance of ionization domains are found to lead at smaller fields to low-frequency fluctuations of the current through the crystal. The uniform glow of the crystal over its volume observed at intermediate fields of 10^3-10^4 V/cm can be ascribed to the same processes that occur in ionization domains, except that the quantitative relationship between these processes is different.

In 1971 we discovered a new phenomenon in sodium zincogermanate crystals [1]: Application of a dc voltage of a few hundred volts results in the appearance at the cathode of a bright green spot, which detaches itself from the cathode and moves towards the anode with a velocity of the order of 10^{-2} cm/sec. In [2] we proposed an explanation of this phenomenon and performed some tentative calculations which confirmed the plausibility and self-consistency of our ideas. The essence of our explanation is that the external electric field produces in the crystal a small (compared with the dimensions of the crystal) region of very strong field in the form of an electrical double layer oriented with the minus towards the cathode. The layers of positive and negative charge can be produced by electrons localized on deep levels; if these levels are sufficiently deep, a charged region of this sort can move through the crystal only very slowly. Electrons traversing this double layer will be accelerated within it, and will impact-ionize all centers located within its confines. Under certain conditions, the layers of both positive and negative charge can be produced and sustained through this ionization. The glow is produced as a result of the recombination of the holes generated by the impact ionization of the crystal lattice.

According to our ideas, the main reason for the appearance of a region of strong field (domain) is impact ionization. We therefore called these domains "ionization domains."

An analysis showed that known electrical instabilities cannot account for the phenomenon discovered by us. Known electrical instabilities arise on account of a random fluctuation, which leads to the appearance of a negative differential conductivity and a strong-field region (electric domain) or a current filament only under certain conditions; no such fluctuation is required, however, for an ionization domain to be produced. Observations and rough estimates soon convinced us that this phenomenon was not fluctuational in character.

In the experiments we observe glowing spots of diameter around 0.5 mm. In external form they recall neither electric domains, which encompass the entire cross section of the crystal, nor current filaments, which extend from the cathode to the anode.

The cross-sectional area of the moving glowing spots in the investigated crystals is several orders of magnitude less than that of the crystal, and in this they are somewhat similar to current filaments. However, unlike current filaments, the luminescence is nonuniformly distributed along the length of the crystal: a sharp maximum of brightness is observed in the vicinity of the spot, a region of weaker and gradually decaying luminescence extending from the spot towards the cathode. A brightness distribution of this sort recalls electric domains. The fact that the dimensions of the spot are limited both with respect to the cross-sectional area of the crystal and its length indicates that the observed phenomenon is of a different nature.

A criterion for the appearance of current filaments or electric domains, i.e., for the appearance of all forms of fluctuation-based electrical instabilities in semiconductors, is that the dynamic volt−ampere characteristic of the semiconductor have a falling region, i.e., a region of negative differential conductivity. The existence of a falling region on the dynamic volt−ampere characteristic means that the static volt−ampere characteristic must have either horizontal or vertical regions.

The volt−ampere characteristic of Mn-activated sodium zincogermanate crystals taken when a glowing spot is present is a smooth curve, and has neither a horizontal nor a vertical region (Fig. 1). On the basis of the above criterion, we can take this to mean that the dynamic volt−ampere characteristic does not have a falling region. Accordingly, fluctuations cannot play the main role in the formation of the spot.

It may thus be taken as established not only that the observed phenomenon has nothing to do with the known forms of electrical instabilities, but also that it is quite unconnected with instability.

Two forms of electroluminescence are known in which the excitation mechanism does not involve instability, namely, injection and prebreakdown electroluminescence.

Let us consider, first of all, whether the injection mechanism of excitation obtain in sodium zincogermanate crystals. The crystals investigated by us were n-type; injection electroluminescence should thus be observed at the anode, as it is from the anode that minority carriers, holes, can enter. In all cases without exception, however, the glowing spots were produced at the cathode. The possibility that an injection excitation mechanism is responsible for the observed phenomenon is thus ruled out.

Prebreakdown electroluminescence may, generally speaking, be recombinational or nonrecombinational in character (in the latter case the luminescing centers are produced

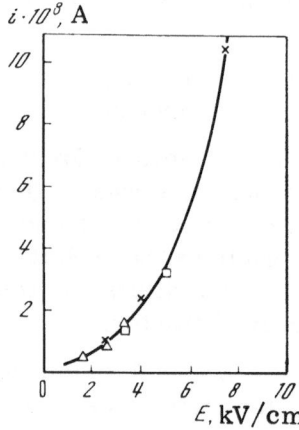

Fig. 1. Volt−ampere characteristic of NaZnGeO$_4$-Mn crystals. The different points refer to different samples.

through impact excitation). Experiments showed that the investigated crystals luminesce with a long afterglow (of the order of 10 min). This indicates that the luminescence is recombinational in character. The excitation mechanism thus involves ionization of the main lattice or of certain levels in the forbidden gap.

Sodium zincogermanate crystals belong to the class of wide bandgap semiconductors. The width of the forbidden band of our crystals was measured in [3] by the electron reflection method; it was found to be 3.94 eV.

The emission spectrum was found to peak at $h\nu = 2.36$ eV [4]; accordingly, radiative recombination occurs on levels spaced from the corresponding band by not less than this amount.

Thermal energy is thus, of course, insufficient to ionize these levels, and in our experiments there was no illumination. Only one possibility remains: ionization is by hot carriers in a region of strong field, i.e., in an electric domain, but one whose production does not, however, involve a random fluctuation.

As our crystals were n-type, it follows that these hot carriers must be electrons.

According to our theory, electrons are accelerated in the high-field region to energies sufficiently high that they can ionize not only any center, but the crystal lattice as well, as a result of which free holes are produced in the crystal. Luminescence occurs in that part of the crystal into which these holes move, i.e., between the cathode and the strong-field region.

Our previous papers [2-5] lacked adequate experimental confirmation of our theory as the properties of the glowing spots had still to be studied in detail. We were convinced only that the theory was free from internal contradictions. In this connection, we estimated some quantities entering into the theory and found that, for reasonable values of these parameters, the factual existence of moving glowing spots could be explained. Of these quantities the most important are:

1) The electric field intensity E in the region of the spot
2) The energy of the electrons, and the manner in which the electrons are distributed over energy in the strong-field region
3) The depth of donor levels which, when ionized, lead to the formation of a layer of positive space charge

To estimate the field intensity E, we found the approximate density ρ of the positive space charge. This charge, in accordance with our hypotheses, is produced by ionized donors, the concentration N_D of which was estimated in order of magnitude in two independent ways from experiment via indirect data. The donor is assumed to be Mn on an Na site. An analysis of the photoluminescent spectra of the investigated Mn-activated crystals showed [4] that the donor concentration was of the order of 10^{18} cm^{-3}. A value of N_D of this same order of magnitude was also found from an analysis of the ESR spectra of the crystals [2, 4].

The potential difference V_0 that could be concentrated in the double layer of the spot was calculated in [2] allowing for the fact that the diameter of the glowing spot is much less than the transverse dimensions of the crystal. It was found that the potential distribution along the symmetry axis of the double layer contains extrema, which vanish if

$$V_0 \leqslant Va/b,$$

where a is the diameter of the strong-field region, b is the distance between the electrodes, and V is the potential difference applied to the crystal. These calculations showed that the electric field intensity in the strong-field regions reaches $\sim 10^6$ V/cm.

The electron energy distribution in the strong-field region, which affects the intensities of the various processes, is of special importance in the theory. This question was studied in [2], where we made the corresponding estimates. The calculations showed that, in the strong-field region, the electrons attain a limiting energy of around 7.4 eV and a minimum energy of around 1.3 eV, and that they are uniformly distributed over energy in this range of energies in the first approximation, as they only acquire energy here and hardly expend any.

We used these estimates to calculate [2, 5] the cross section for donor ionization; it worked out to be $3 \cdot 10^{-14}$ cm^2. In this calculation we utilized the experimentally measured mean velocity of motion of the glowing spot (v = $4 \cdot 10^{-2}$ cm/sec) and the mean value of the current through the crystal (10 μA). The fact that such a reasonable value was obtained for the donor ionization cross section we regarded as one of the confirmations of our ideas.

Another important assumption made in the theory is that the crystal has deep donor levels and relatively shallow acceptor levels; electrons can be freed from donors only through impact ionization (and not by the tunnel effect), whereas, by contrast, the freeing of holes from acceptors by the tunnel effect is highly probable. Our estimates indicate that the depth of the donor level must lie in the range from 0.7 to 1.3 eV. Such a level, of depth 1.0 eV, was discovered experimentally in [5, 6]. This we regard as a second confirmation of our theory.

In the work described here we investigated the quantitative characteristics of individual spots and the background luminescence in the absence of spots. Also, the theory previously proposed was further developed and checked.

First of all, we wished to check the proposition that the formation of a domain is accompanied by a redistribution of the current, and that the current through the crystal is mainly determined by the current through the domain. Concurrently with this, we checked the assertion that it is precisely this current that is responsible for the very existence of the domain and its motion from the cathode to the anode. To this end, we measured the dependence on time of the "instantaneous" values of i and v, respectively the current through the crystal and the velocity of motion of the glowing spot, and also the relationship between these quantities [7, 8]. According to our theory, this relationship is a very simple one:

$$v \sim i. \tag{1}$$

Oscillograms of the current can be used to make an independent check on this prediction of the theory. Indeed, if it is correct, we see that the distance x traversed by the spot must be proportional to the area under the oscillogram of the current in the corresponding interval of time:

$$x = \int_0^t v \, dt \sim \int_0^t i \, dt. \tag{2}$$

The experimental method of observing the instantaneous values of current and brightness is simple and is described in [5, 8]; accordingly, we shall not concern ourselves with it here. Typical oscillograms of the current and brightness during the motion of the spot are shown in Fig. 2.

In all cases, when marks corresponding to the moment the spot reached the middle of the crystal were made on the oscillograms, equality (2) was obeyed to within not worse than 20%. (Better agreement cannot be counted upon, as the position of the spot was determined by eye.) The time taken by the spot to traverse the first half of the crystal was three times

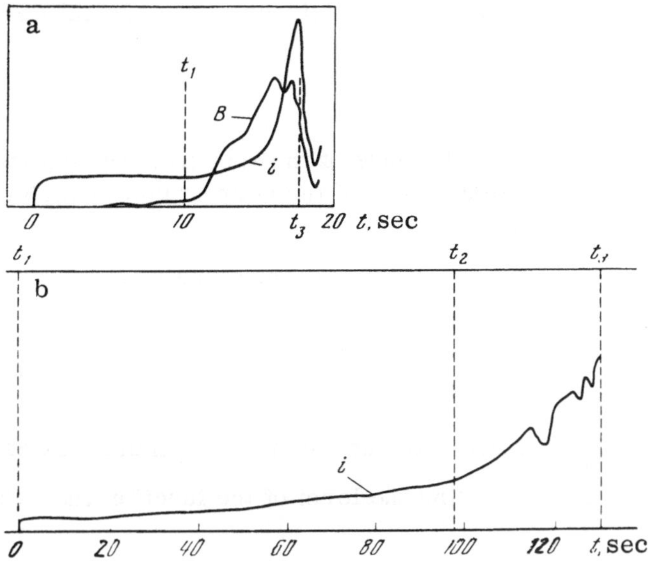

Fig. 2. Oscillograms of current i and brightness B during motion of glowing spot. (a) Rapidly moving spot; (b) slowly moving spot. Time t_1 is moment at which spot appears; t_2 is moment at which spot reaches middle of crystal; t_3 is moment at which spot reaches anode.

greater than the time to traverse the second half, i.e., the spot moves with an acceleration (Fig. 2b).

The proportionality between the instantaneous values of the current through the crystal and the velocity of motion of the spot predicted by the theory of ionization domains can thus be taken as experimentally verified. This is yet further evidence that the measured current passes almost entirely through the spot, i.e., is very nonuniformly distributed within the crystal. Nonetheless, we checked this proposition in yet another way, based on the fact that the current through the crystal increases the closer the spot comes to the anode. This experimental fact is readily explained within the framework of our theory: intense ionization of the crystal lattice occurs in the region of the domain, as a result of which an excited region of enhanced conductivity remains behind it. It is recombination in this region that gives rise to the glow observed between the spot and the cathode. According to the theory, almost all the current flowing through the crystal passes through the ionization domain; accordingly, despite its small cross-sectional area, the conducting channel produced behind the domain* significantly decreases the total resistance of the crystal. As a result, the current should increase the closer the spot comes to the anode, which is indeed seen to be the case in the oscillograms of Fig. 2.

We shall assume that the conductivity of the conducting region is constant along its entire length† and that the diameter of the current tube increases only insignificantly between the spot

---

* In our crystals, such a channel differs from a current filament not only in the mechanism whereby it is produced, but also in the absolute value of the mean current through the crystal (of the order of a microamp) at which is appears.

† The resistivity of the excited part of the crystal r_x increases, of course, with the passage of time, approaching the resistivity of the unexcited part r. This occurs very slowly, however, which can be judged from the rate (more exactly, the slowness) of decay of the luminescence, this being determined by the same recombination processes responsible for the fall in the conductivity. By our observations, the afterglow lasts for minutes. Accordingly, if the time for which the spot moves is small, of the order of a few seconds, r_x can be regarded as independent of time and so also of the coordinate x. In the present case we assume that r_x is constant. In the next case to be considered, however, that of a slowly moving spot, the increase of r_x will be taken into account.

and the anode. The resistance of the crystal can then be written

$$R = \frac{1}{S} \left[x r_x + (d - x) r \right].$$ (3)

Here x is the distance of the spot from the cathode, d is the length of the crystal, and S is the cross-sectional area of the current tube, which can be regarded as equal to the area of the glowing spot.

The current through the domain is

$$i = \frac{VS}{rd - (r - r_x) x}.$$ (4)

As $r_x < r$, we see that the current i increases as the spot approaches the anode.

Let us find the form of the function i(t). Utilizing relationship (1), we have

$$\frac{dx}{dt} = \frac{k}{rd - (r - r_x) x},$$

where k is a coefficient of proportionality. Solving this equation for x(t) and inserting the result into (4) gives

$$i(t) = \frac{VS}{\sqrt{(rd)^2 - 2(r - r_x) kt}}.$$ (5)

It can be seen from this expression that $1/i^2(t)$ depends linearly on time. A dependence of this sort is indeed observed experimentally for rapidly moving spots (Fig. 3, curve a).

In the case of a slowly moving spot, however, the dependence of r_x on time, and so also on the coordinate x, can no longer be neglected. The increase of r_x as the conductivity decays leads to a less rapidly increasing current and thus to a smaller (than in the first case) acceleration of the spot.

As a result, the points on the plot of $1/i^2 = f(t)$ will be displaced more and more to the right (in the direction of larger t) away from the straight line defined by the initial part of this plot. In Fig. 3 this situation is depicted by curve b. It can be seen that this curve is concave

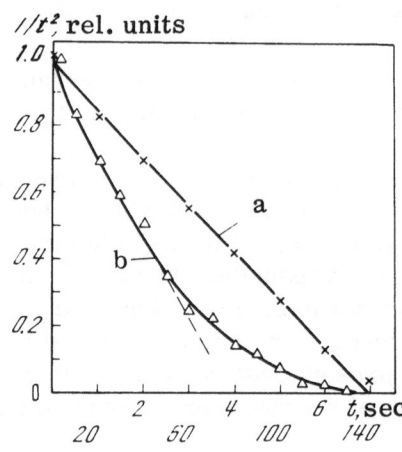

Fig. 3. Plots of $1/i^2(t)$. (a) Rapidly moving spot (upper time scale); (b) slowly moving sport (lower scale). The corresponding oscillograms are shown in Fig. 2 (a and b).

downwards. The spot here was moving about 20 times more slowly than in case a, with the result that the decay of the conductivity can manifest itself fully.

In this manner, our theory explains why the current increases as the spot approaches the anode, a feature which cannot be accounted for by theories based on electrical instabilities, since, for ordinary electric domains, the current turns out to be constant during the entire lifetime of the domain.

The check on the theory of ionization domains cannot end here, however, since the postulates made in the theory concerning the behavior of electrons within a domain must also apply, to some extent, to electrons outside the domain. In particular, we assumed that an electron possessing a sufficiently large energy is scarcely deflected from its path as a result of a collision with a charged impurity. More precisely, we assumed that this deflection does not exceed 90°, i.e., that the electrons are not turned backwards as a consequence of a scattering event.

This requirement has a very profound effect on the electron energy distribution; if the electrons were back-scattered, they would start to move against the field force acting on them and would continue to lose their energy even after the scattering event, giving it up to the electric field. In this manner, they can lose almost all their energy. Under normal conditions scattering through large angles ("backwards") is the main reason for the weak heating of the electrons in the field.

In an electric field of the order of 10^6 V/cm and above, which, according to our calculations, exists in an ionization domain, an electron has time to acquire an energy of almost 1 eV even in a distance equal to one mean free path without the field. Indeed, this was why we assumed back-scattering to play only a small role.

However, even in weaker fields, and, indeed, in the absence of any field at all, in a state of thermodynamic equilibrium, a semiconductor will always contain a certain (albeit very small) number of electrons of sufficient energy, and, consequently, momentum, such that they are scarcely deflected from their path as a result of interaction with a scattering center or phonons.

These electrons, according to our ideas, behave in an essentially different manner in an external electric field. Consequently, having utilized this feature once (to explain the properties of the domains), we must, for consistency, establish what sort of role these electrons play outside the domain, where there is also an electric field, albeit two or three orders of magnitude weaker.

Let us consider this question in more detail. The electron mean free path is determined, as is well known, by two main processes: scattering on charged centers, and scattering on phonons. As mentioned previously, the investigated crystals were heavily doped and compensated. The concentration of charged centers was thus large, and such centers played the main role in the scattering in these crystals.

The effective cross section for this scattering process, S_s, can be estimated by calculating the distance in which the kinetic energy of an electron becomes comparable with its potential energy in the field of the scattering center. Taking this as the scattering radius, we obtain:

$$S_s = \pi e^4/\varepsilon^2 W^2, \tag{6}$$

where ε is the dielectric constant of the crystal, and e and W are respectively the electronic charge and the energy of an electron above the "bottom" of the conduction band. It can be seen from this expression that the effective cross section decreases rapidly with increasing W.

The quantity S_s characterizes scattering in all directions. We are interested in the effective cross section for scattering "backwards," which is, of course, several times smaller. It we employ Rutherford's formula for the scattering of charged particles by a repulsive center and integrate it over all angles corresponding to scattering "backwards," we obtain an expression that differs from (6) only by a coefficient of 1/4:

$$S_{s\,\text{back}} = \pi e^4/4\varepsilon^2 W^2. \tag{7}$$

This is the expression which we shall utilize below.

Let us work out what sort of path, on average, will be traversed by an electron in a field of intensity E. Interaction with phonons will be neglected for the moment. We suppose that 1 cm^3 of the crystal contains N scattering centers with an effective scattering cross section described by Eq. (7). The distance x traversed, on average, by an electron before it experiences a scattering event is then given by

$$NS_{s\,\text{back}}\, x = 1. \tag{8}$$

However, if this motion takes place in an electric field E, the kinetic energy of the electron increases along its path from W_0 to a value

$$W = W_0 + eEx. \tag{9}$$

(The x axis points in the direction in which the electron moves under the action of the electric field.) We insert W into (7) and replace in (8) multiplication by x by integration with respect to dx, since $S_{s\,\text{back}}$ is now itself dependent on x. We obtain, as a result,

$$\frac{\pi e^4 N}{4\varepsilon^2} \int_0^x \frac{dx}{(W_0 + eEx)^2} = 1, \tag{10}$$

which, after integration and some straightforward algebra, gives the following expression for x:

$$x = \frac{4\varepsilon^2 W_0^2}{\pi N e^4}\, \frac{1}{1 - 4\varepsilon^2 W_0 E/\pi N e^3}. \tag{11}$$

It can be seen that the denominator of this expression can become equal to zero or even negative for certain sufficiently large W_0 and E and not too large N. This means that the electron is not scattered at all by the charged centers, since the further it moves the greater the energy it acquires and the less it feels the Coulomb field of the center.

The above discussion is not, of course, a rigorous one, as the effective scattering cross section cannot be reduced indefinitely. It shows, however, that in an electric field the electron mean free path can be much greater than in zero field, especially if the electron has acquired, for some reason or other, additional energy.

Let us now consider the interaction of electrons with phonons, which we have hitherto neglected. Energy exchange with phonons can lead, at worst, to a loss of energy equal to the energy of the created phonon. If, between two phonon-creation events, the electron passes through a potential difference that is more than sufficient to make up this loss, then on average the electron will still gain energy, although somewhat more slowly than if energy exchange with phonons had been absent. Exchange of momentum can act in the same way as scattering on charged centers. The probability of creation of a phonon is then greatest precisely for those electrons that experience a large change of momentum, since each created phonon carries away a large momentum. We shall thus assume that phonons are created only by those electrons which are scattered backwards, and which, consequently, we need no longer consider.*

Scattering is greatest on longitudinal optical (LO) phonons, as they create fluctuations of electric charge. These fluctuations act, essentially, in the same manner as charged centers. Accordingly, high-energy electrons are likewise scattered on them less than low-energy electrons. We may therefore assume, in the first approximation, that scattering on phonons is equivalent, in a sense, to a certain increase in the concentration of the charged impurity.

We now go on to make a quantitative estimate. According to our data, in the investigated crystals $N_D - N_A = 5 \cdot 10^{17}$ cm^{-3} for a degree of compensation of approximately 0.8. This means that the total number of charged scattering centers is $4 \cdot 10^{18}$ cm^{-3}. Scattering on impurities usually predominates in heavily doped semiconductors. We shall thus assume that the scattering on phonons is equivalent to a certain increase in the concentration of the scattering centers. For definiteness, we take increase to be 25%. (In fact, scattering on phonons is probably even less in our case.) As before, we take the dielectric constant of the crystal to be 6. Inserting these values into (11) gives:

$$x = \frac{4.5 W_0^{-2} \cdot 10^{-4}}{1 - 4.5 \cdot 10^{-4} W_0 E}, \tag{12}$$

where the electron energy W_0 is expressed in electron volts, the electric field intensity E is in V/cm, and distance x is in centimeters.

In our samples, at a mean field intensity of $4-5 \cdot 10^3$ V/cm, the whole volume commenced to glow brightly (previously the glow was weak) and the moving spots associated with the ionization domains could no longer be distinguished on the background of this glow. At this sort of field intensity, we see from (12) that the initial energy of an electron must not be less than 0.56 eV if it is to acquire, in an unimpeded manner, an energy sufficient to ionize the crystal lattice. An electron with a smaller initial energy can acquire sufficient energy to ionize the lattice only if, by chance, it traverses without a scattering event a distance considerably greater than the mean free path. For example, for an initial energy of half the above amount (i.e., 0.28 eV), an electron, on traveling a distance equal to a mean free path (allowing for the dependence of the effective scattering cross section on electron energy), acquires a further 0.28 eV. Accordingly, if by chance it experiences no scattering event over this distance (the probability of this equals e^{-1}), it will thereafter be no different from an electron with an initial energy of 0.56 eV and may, consequently, accumulate an energy sufficient to ionize the crystal lattice.

Let us consider whether thermal motion is capable of providing the number of high-energy electrons required to maintain the experimentally observed glow brightness, which amounts, in order of magnitude, to a few apostilbs. When allowance is made for the thickness of the investigated crystals in the direction of observation (2 mm) and for the energy of the emitted quanta (2.4 eV), we find that this corresponds to the creation of around 10^{12}-10^{13}

---

*In [2] we estimated the interaction with phonons in a different way, allowing only for energy exchange (in an exaggerated form, as we assumed a new phonon to be created in each interaction event) and neglecting the possibility of back-scattering. This method gives an averaged estimate, and is suitable only for very strong fields, when fluctuations of the mean free path and the scattering angle are insignificant. The method employed in the present paper to estimate interaction with phonons seems to us to be more exact.

quanta/cm^3 · sec. The current through the crystal under these conditions was of the order of 1 μA (at a field intensity of 4 · 10^3 V/cm).*

When allowance is made for the effective electron mass (according to our data, 0.15 m$_e$) and for the concentration of scattering impurities (4 · 10^{18} cm^{-3} plus 1 · 10^{18} cm^{-3} to allow for interaction with phonons), we find that the mobility of electrons in these crystals equals approximately 80 cm^2/V · sec. For this value of the mobility and for the values of the current and voltage measured in the experiments, the free-electron concentration in the crystal works out to be 2 · 10^{18} cm^{-3}. The electron concentration at a height of 0.56 eV above the bottom of the conduction band will be much less. On working out the corresponding Boltzmann factor for the temperature of the experiment (330°K), we find that the electron concentration at a height of 0.56 eV will amount in all to only 0.42 cm^{-3}. Further, of them only one-sixth will move in the direction of the field (the remainder will move in the opposite direction, or sideways). In this manner, only one electron in 7 cm^3 will be able to acquire, unimpeded, an energy sufficient to ionize the lattice!

It would appear to be obvious that one electron in 7 cm^3 of a crystal of volume around 0.02 cm^3 will be incapable of causing a glow of any noticeable brightness. In fact, however, it is the opposite that is the case. An electron raised by thermal fluctuations to such a high level in the band lives there for only a very short time. The time required for total thermalization of electrons was estimated in [9] as 10^{-11}-10^{-9} sec; the time to drop a few hundredths of an electron volt is, of course, one or two orders of magnitude smaller than this. It follows that even this negligible concentration of electrons at a height of 0.56 eV can be maintained (in the absence of a field) only by the arrival at that level per second of 10^{10}-10^{11} electrons/cm^3. (The 1/6 has been taken into account here.) In a field almost all the electrons can freely acquire an energy sufficient to ionize the lattice, with the result that the creation of at least 10^{10} quanta/cm^3 · sec is assured, which is no longer such a catastrophic number of times smaller than is observed experimentally. If we remember, however, that electrons having half this energy (i.e., 0.28 eV) are capable, with a probability smaller by only a factor of e, of acquiring an energy sufficient to ionize the lattice, it turns out that the number of electrons capable of ionizing the lattice is almost four orders of magnitude greater than the above figure. This is now more than enough to give the number of emitted quanta observed experimentally, even if the luminescence quantum yield is only 10%.

The number of electrons capable of reaching energies sufficient to ionize the crystal lattice depends very sharply on the magnitude of the applied electric field. There are two reasons for this: firstly, the initial energy of an electron for which it can subsequently be freely accelerated increases with decreasing field; secondly, there is an increase in the distance that an electron must travel without a single random collision in order that it acquire in the field this initial energy. Thus, if the applied field is decreased from 4 · 10^3 to 2 · 10^3 V/cm, ie., in all by a factor of 2, the brightness should decrease by more than four orders of magnitude.

Experimentally, the brightness does indeed fall rapidly with decreasing applied voltage, but not just quite so sharply as this. This is probably because, besides thermal motion, a supply of high-energy electrons capable of being accelerated in the field may come from the near-cathode layer if the latter contains a layer of space charge giving rise to an electric field stronger than the mean field in the crystal. It is true that the near-cathode layer can supply high-energy electrons only to the immediately adjacent part of the crystal. The glow extends, however, more or less uniformly over the whole crystal. This feature can also be explained within the framework of the present theory.

---

* Different samples differed greatly both in brightness and in current. Accordingly, all numerical data cited here and below must be regarded as very tentative.

For this purpose we require to consider the processes occurring after the electron has acquired an energy equal to the ionization threshold for the crystal lattice. For definiteness, we shall consider the processes occurring in a field of intensity $1.6 \cdot 10^3$ V/cm, which is close to the lower limit for the appearance of a uniform glow. As was shown previously, once it has acquired an energy equal to the lattice ionization threshold, the electron continues to move in the electric field for a certain time, accumulating additional energy. This comes about because the effective ionization cross section has a finite value, being even close to zero near the ionization threshold. Calculation shows that if, after the threshold energy has been attained, the effective ionization cross section increases according to a quadratic law and reaches, far from threshold, a value of 10^{-15} cm^2, then, prior to the act of ionizing the lattice, the electron manages to accumulate 0.2-0.3 eV above the threshold energy.

If we assume, as before, that the ionization threshold is 1.5 times the width of the forbidden band, we find that after an ionizing event an energy of 2.2-2.3 eV must be distributed among the three charges (two electrons and one hole). If the energy of the ionizing electron is exactly equal to the threshold energy and if the electron and hole effective mass is the same, the excess energy is uniformly distributed among the charges. If, however, the ionizing electron has an energy greater than threshold, momentum conservation implies that the excess energy is distributed nonuniformly among the three charges. If, after the ionizing event, all three charges have a momentum directed in the same direction as the momentum of the electron causing the ionization, and if two of the three particles have the same energy, then the problem has two solutions: (1) two particles have an energy of 1 eV each, and the third an energy slightly greater than 0.2 eV; (2) two particles have an energy slightly less than 0.5 eV, and the third an energy of almost 1.4 eV. It is not possible to say at the moment which of these two cases occurs in experiment, or whether the particles with the same energy are the two electrons, as particles of the same sign, or an electron and a hole, as the newly created particles. For our purposes, however, this is not important. The significant thing is that, in either case, there remains at least one electron with an energy of the order of 1 eV.

We return now to the explanation of why the glow is uniform over the volume at small electric field intensities. As we did not take any steps to make the contacts with the crystal ohmic or at least antibarrier, and as a volume glow was observed for the most varied types of contacts (indium—gallium paste, Aquadag, special contact enamel, rubbing metal electrodes), a certain potential barrier probably always existed across the contact, through which the electrons passed as they moved from the contact into the crystal. As is well known, this sort of barrier is hard to avoid. As the crystals had n-type conductivity, this barrier must lead to the formation of a layer of positive space charge near the cathode. If a voltage of even 6-7 V is dropped across this layer (out of the several hundreds applied to the crystal), then, clearly, high-energy electrons, albeit in small numbers, must emerge from it.

Calculation by Eq. (11) shows that, in a field of $1.6 \cdot 10^3$ V/cm, electrons with an energy exceeding 1.4 eV will be freely accelerated. If, from the near-cathode layer of space charge, where the field is very much greater, electrons with this energy or greater emerge, then they will be accelerated in the relatively weak field occurring in the rest of the crystal and will reach an energy sufficient to ionize the crystal lattice. If case (2) above arises as a result of an ionizing event, i.e., if, after ionization, one of the electrons has an energy of almost 1.4 eV, then it will be freely acclerated* and may again acquire an energy sufficient to ionize the lat-

---

* Under the given conditions, this electron will have an energy of 1.37 eV. An electron will have an energy of 1.37 eV. An electron of this energy will be freely accelerated in a field not of $1.6 \cdot 10^3$ but of $1.65 \cdot 10^3$ V/cm. This difference will be ignored, as all the calculations are of a tentative character.

tice. This may be repeated many times, until the electron passes right through the crystal and comes out at the anode.

If, however, the ionizing event results in case (1) above, two possibilities arise: (a) if only one electron has an energy of 1 eV, the glow will decay rapidly with increasing distance from the cathode, since the probability that this electron will acquire an energy sufficient to ionize the lattice is appreciably less than unity; (b) if both electrons have an energy of 1 eV, then, since the probability of attaining a high energy is still greater than 1/2 for each of them, it follows that, on average, at least one of them will be able to keep the process going. This implies that the glow will be distributed over the entire crystal. Which of these two possible variants, (1a) or (1b), occurs in reality cannot be said at the moment.

Let us now estimate the possible intensity of the luminescence. To this end, we require to consider the conditions in the near-cathode layer of space charge. The maximum charge density in this layer is $N_D - N_A$, i.e., according to our data, $5 \cdot 10^{17}$ cm^{-3}. However, when a flux of electrons passes through this region, some of the electrons may be trapped by ionized donors, thereby reducing the net positive charge. They will be able to free themselves as a result of thermal fluctuations and also as a result of events involving those of their "compatriots" that manage to acquire sufficient energy. Although, as mentioned previously, the depth of the donor levels in these crystals amounts to around 1 eV, the number of electrons possessing an energy of this amount or greater in the near-cathode region will be quite large, since calculation shows that the field in this region is of the order of 10^5 V/cm.

The probability of an electron becoming localized on an attractive center is, generally speaking, less than the probability of its being scattered by such a center, since, if it becomes localized, it must give up its kinetic energy to something or other (e.g., on the creation of several phonons). We thus assume that the only electrons that can become localized are those that manage to acquire an additional energy W_1 which is not greater than the phonon energy. The effective capture cross section in this case we shall assume to equal the effective thermal-electron scattering cross section S_0, and we take the capture cross section for electrons of greater energy to be equal to zero.

If N is the concentration of ionized donors, the probability of an electron being trapped by an ionized donor in a distance x is $NS_0 x_1$. The quantity x_1 here is the distance in which the electron acquires, in the field E, an energy W_1, i.e., we set $x_1 = W_1/eE$. The number of scattering events per unit time per unit volume is given by the product $NS_0 x_1 j$, where j is the density of the electron flux. The number of events in which electrons are liberated will equal $\nu S_1 x_0 j p$, where ν is the concentration of neutral donors, S_1 is the effective cross section for liberation by electron impact, x_0 is the thickness of that part of the near-cathode space-charge layer where there are high-energy electrons, and p is the proportion of these electrons in the total flux. The quantity p is unknown, and, moreover, it varies as a function of distance within the layer; estimates made below indicate, however, that it amounts to tens of percent.

Under steady-state conditions, the number of events in which electrons are localized must equal the number of events in which they are liberated, i.e.,

$$NS_0 \, \frac{W_1}{eE} \, j = \nu S_1 x_0 pj, \tag{13}$$

whence

$$\nu = N \, \frac{S_0}{S_1} \, \frac{W_1}{eEx_0 p} \,. \tag{14}$$

If we take the field E as the mean field in the space-charge layer and x_0 as the total thickness of this layer, then the product Ex_0 represents the potential difference dropped across the

layer, V_b. For S_1 we take the value $3 \cdot 10^{-14}$ cm^2, in accordance with our previous paper [2].*
The energy W_1 we take to be 0.03 eV, since, as indicated above, it must be close to the phonon
energy. We then have that $\nu = N \cdot 0.03/p$. This means that under the considered conditions
only about 10% of the donors remain in the unionized state.

However, under equilibrium conditions, in zero field, there are in all 15-20% un-ionized
donors, since the degree of compensation of our crystals is not less than 0.8 [2]. It follows
from this that, under the given conditions, the space-charge density in the near-cathode region
is appreciably less than the maximum possible.† We shall assume that, instead of being
$5 \cdot 10^{17}$ cm^{-3}, it amounts to $2 \cdot 10^{17}$ cm^{-3} over the entire thickness of the space-charge layer
(which in itself, of course, is a rough approximation).

Let us now calculate the probability of the electron traversing the space-charge layer
without being scattered backwards. As before, the reduction of the effective scattering cross
section with increasing electron energy will be taken into account. On top of this, we have to
allow for the fact that the electric field in the near-cathode space-charge layer is distributed
nonuniformly, resulting in a more complex formula for $S(x)$:

$$S(x) = \frac{S_0 W_0^2}{\left(W_0^2 + \int\limits_0^x eE dx\right)^2}. \tag{15}$$

Here S_0 is the effective scattering cross section corresponding to the initial electron energy W_0.

After some protracted algebra, we obtain

$$\int\limits_0^{x_0} NS(x)\,dx = \frac{NS_0 x_0}{A+1}\left(1 + \frac{1}{\sqrt{A^2+A}}\ln\frac{\sqrt{1+1/A}+1}{\sqrt{1+1/A}-1}\right), \tag{16}$$

where

$$A = ev/W_0.$$

The initial electron energy W_0 we take to be $^1/_2 kT$, since, as far as acceleration is concerned, it is the component of electron velocity along the field that is important.

In our case $A \gg 1$. Consequently, Eq. (16) can be simplified greatly by discarding the
logarithmic term in the brackets and the "1" in the denominator of the quantity in front of the
bracket. Inserting numerical values into the resultant expression, we find that, for the concentration of scattering centers which, by our data, occurs in the volume of the crystal, the
quantity $\int\limits_0^{x_0} NS dx \approx 0.5$. Near the surface, however, the scattering centers are usually larger,
due to the considerable imperfection of layers near the surface and also to possible diffusion
into this region of material from the electrodes.

---

* Thermal liberation of electrons is neglected at the moment, as in the present case this pro-
ceeds very much more slowly.
† It should be remembered that, as the field in the near-cathode layer increases, the probability
of an electron being trapped (liberated) decreases (increases). The quantities E, x_0, and p
entering into the denominator in (14) are thereby increased. The concentration of un-ionized
donors thus falls sharply, and the charge density rapidly approaches the limiting value. It
becomes effectively equal to the limiting value for a voltage across the layer of 3-4 times
greater.

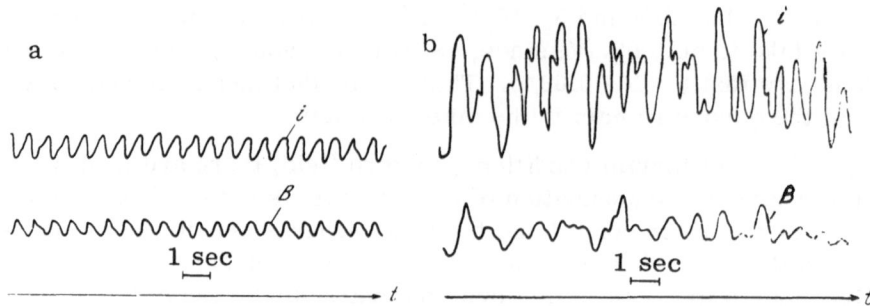

Fig. 4. Oscillograms of brightness B and of current i through crystal for fast processes.

In this manner, we see that more than half of the electrons experience a scattering event as they pass through the near-cathode layer, although a considerable proportion traverse it nonetheless without being scattered. Electrons which have already experienced a scattering event somewhere within the layer, and which have lost their energy as a result, have a much greater chance of being scattered again, as they are in a region of lower field intensity and accumulate energy more slowly. Thus, an electron cropping up with energy W_0 in the middle of the space-charge layer is twice as likely to be scattered in the remaining half of its path as an electron which has only just tunneled through from the electrode and which has, ahead of it, a path in the space-charge region that is twice as great.

In this manner, two fluxes of electrons of electrons of sharply differing energies will emerge from the near-cathode layer: electrons with an energy of a few electron volts that are "picked up" by the electric field in the bulk of the crystal, are accelerated, and impact-ionize the lattice (as was described previously); and electrons which have been scattered, which have lost almost all their energy,* and which are accordingly rapidly thermalized after they emerge from the strong-field region.

It follows from the above that the bulk luminescence of the crystal is due to a few tens of percent of the total number of electrons that pass through it. The current through the crystal was, in this case, more than 10^{-9} Å, i.e., $6 \cdot 10^9$ electrons/sec, while the light flux was of the order of 10^9 quanta/sec. Remembering, further, that each high-energy electron can be accelerated again and again, causing impact ionization each time, it becomes clear that the current passing through the crystal can, in this case too, give rise to the experimentally observed brightness.

There is just one more feature of the bulk luminescence of the investigated crystals that remains to be explained, namely, the fact that the brightness and the current are not constant in time. In some of the investigated samples the oscillations of current and brightness are strictly periodic, with the same frequency, of the order of one or several Hertz (Fig. 4a); in others, these parameters vary in an irregular manner (Fig. 4b). A correlation analysis shows, however, that in the latter case too it is frequencies of this same order of magnitude that predominate. Probably, the latter samples differ from the former only in that different parts of them give rise to oscillations of slightly different frequencies, as a result of which the overall picture is irregular in appearance. Finally, there were samples in which these oscillations were almost imperceptible.

---

*We recall that we have in mind only scattering "backwards," when the electron gives up almost all its energy to the field.

In those cases when the moving ionization domains discussed at the beginning of the present paper were produced, regular oscillations were not observed, not even once, while the irregular oscillations had a somewhat smaller relative amplitude, probably because the main light flux came from the vicinity of the domain.

The irregular variation of the current and brightness can be understood from a more detailed study of those same processes in the near-cathode region that we have just been considering in connection with the uniform luminescence over the volume of the crystal at relatively small voltages, and also from a study of the processes occurring in ionization domains.

As mentioned earlier, the negative component of the electrical double layer is produced through impact deionization of donors by high-energy electrons, as a result of which electrons are transferred from the valence band directly to the donor level. This process begins, of course, at voltages that are still too low for the formation of ionization domains as such. As a result, the space-charge density in part of the near-cathode layer is reduced somewhat, although this plays no role prior to the formation of the negative charge.

The reduction of the space-charge density leads to a reduction of the field intensity, both in the immediate vicinity of the cathode and in the middle of the space charge. Accordingly, on the one hand, tunneling of electrons from the cathode is reduced, and, on the other, the energy of the electrons is reduced, which slows down the deionization of the donors. The space-charge density is thus gradually restored. As soon as it is restored, the process begins all over again.

The processes specified above need not, generally speaking, result in the appearance of an oscillatory mode, as there is another possibility — the establishment of a steady state with a slightly diminished charge density. With a view to establishing what must, in fact, occur in reality, let us consider a simplified model of the near-cathode space-charge layer. Let us suppose that there is concentrated in this layer a voltage not of 6-7 V, but a somewhat greater voltage, of say 10-12 V. Under these conditions electrons with an energy sufficient to deionize donors will now appear in the middle part of the layer.

We divide the space-charge region into three parts: (1) from 0 to x, in which the charge density is a constant, equal to the limiting charge density ρ_0 (in this part the electrons are being accelerated only, although they are still unable to ionize the crystal lattice or to deionize donors); (2) from x_1 to x_2, in which this deionization occurs quite intensively, although it does not prevail over impact ionization of the donors (in this part, the charge density ρ depends both on the coordinate x and on time); (3) from x_2 to the boundary of the space-charge layer x_3, in which the field is so weak that the only electrons capable of being accelerated are those that managed to acquire sufficient energy in parts (1) and (2). We shall assume that the space-charge density in this part is constant and equal to ρ_0, as in the first region, but that the boundary x_3 is movable. The latter requirement is necessary because the total voltage drop across the space-charge layer is determined, on the one hand, by the charge distribution in it, and, on the other, by the difference between the total applied voltage V and the ohmic voltage drop in the bulk of the crystal iR, and these two different physical circumstances must automatically accommodate each other. We may then formally regard the boundaries x_1 and x_2 as stationary (Fig. 5).

Let us construct the appropriate system of kinetic equations. As mentioned previously, electrons pass into the crystal from the cathode by the tunnel mechanism. The tunneling probability is determined by the height and the width of the corresponding barrier. The barrier height does not depend on the applied voltage and is determined by the corresponding heterojunction, while the width decreases with increasing field in the barrier. The only barriers that can be penetrated to any significant extent are those of very small width, of the order of a few tens of angstroms. Accordingly, in the first approximation, the field in the barrier can be taken to be constant and equal to E_m. The density j of the current passing through the barrier

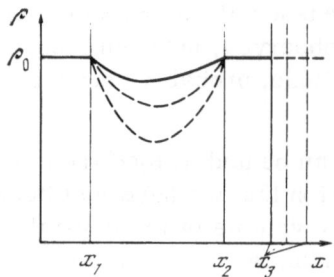

Fig. 5. Charge density distribution in near-cathode layer. The dashed curves show the distribution at other moments of time.

will then equal

$$j = D \exp\left(-\frac{4\sqrt{m^* u^3}}{3heE_m}\right),$$ (17)

where u is the barrier height, m^* and e are the electronic effective mass and charge, h is Planck's constant, and D is a constant the value of which depends on the properties of the junction.

The variation in time of the charge density ρ in part (2) of the space-charge layer will be described by the equation:

$$\partial\rho/\partial t = -S_{DI}(N_D - \rho_0 + \rho)j + S_I(\rho_0 - \rho)j.$$ (18)

Here S_{DI} and S_I are the effective cross sections for donor de-ionization and ionization respectively. They depend on how the electrons traversing this region are distributed over energy, i.e., on the coordinate x. We assume that they are independent of time in the first approximation. The difference $\rho_0 - \rho$ is the concentration of neutral donors.

The maximum electric field E_m is found simply by integrating over the entire space-charge region (we neglect the field in the bulk of the crystal):

$$E_m = \frac{4\pi\rho_0}{\varepsilon}(x_1 + x_3 - x_2) + \frac{4\pi}{\varepsilon}\int_{x_1}^{x_2}\rho(x)\,dx.$$ (19)

Finally, we obtain a fourth equation by calculating, in two independent ways, the voltage V_b dropped across the space-charge layer (this was mentioned previously in connection with the mobility of boundary x_3). The function $E(x)$ which we have to integrate in order to obtain V_b can be found in the same manner as E_m except that the integration has to be carried out not from x_3 to 0, but from x_3 to x. Equating the expressions for V_b obtained by the two methods gives

$$V_0 - jS_{\text{cryst}}\,Re = \frac{2\pi\rho_0}{\varepsilon}(x_1^2 + x_3^2 - x_2^2) + \frac{4\pi}{\varepsilon}\left[\int_{x_1}^{x_2}\!\!\int_{x}^{x_2}\rho(u)\,du\,dx + x_1\int_{x_1}^{x_3}\rho(x)\,dx\right].$$ (20)

Here S_{cryst} is the cross-sectional area of the crystal.

These four equations contain the four unknown functions: j(t), E_m(t), x_3(t), and ρ(x, t).

First of all, let us consider whether this system of equations permits of a stationary solution. Setting $\partial\rho/\partial t = 0$, we find ρ from (18) and insert the resulting expression into expressions (19) and (20).

Equation (20) will then contain only the one unknown, j, which can thus be found. But our system of equations yields a second equation for j, obtained by inserting into Eq. (17) the quantity E_m from Eq. (19), in which ρ has already been expressed in terms of j. Clearly, these two equations will, generally speaking, be incompatible, as they contain different mutually independent numerical parameters. It follows that, generally speaking, system (17)-(20) does not possess a stationary solution.

Let us now establish the nature of the nonstationary solution. It can be seen from (19) and (17) that the greater $\rho(x)$, the greater E_m, and so also j. It thus follows from Eq. (18) that if ρ is close to ρ_0 (it cannot be greater than ρ_0), then $\partial\rho/\partial t < 0$. If, however, ρ is small, then, if the ratio S_{DI}/S_I is sufficiently small, it may transpire that $\partial\rho/\partial t > 0$. As can be seen from (18), the value of this ratio at which ρ is still able to increase depends on the degree of compensation of the crystal. (We recall that $\rho_0 = N_D - N_A$; as ρ is taken to be nonnegative, it follows that the smallest value it can have is zero.) Since the degree of compensation of our crystals is 0.8, it follows that ρ can increase provided $S_{DI} < 0.25 S_I$ (a degree of compensation of 0.85 would require $S_{DI} < 0.17 S_I$).

The relationship between S_{DI} and S_I depends on the electron energy W. Both these cross sections are nonzero only if W exceeds a certain threshold energy W_{th}. As W increases above this value, the cross sections increase, reach a maximum, and then begin to decrease for $W \gg W_{th}$. As the threshold energy for the ionization of donors (1 eV) is one-third of the threshold energy for de-ionization, it follows that quite a broad range of electron energies exists in which $S_{DI} < 0.2 S_I$. This range starts at 1 eV and ends somewhere after 3, or, more likely, after about 4-5 eV. The voltage across the barrier and the boundaries x_1 and x_2 of the middle part of the space-charge layer were chosen with this taken into account. Integration of (19) with respect to time gives

$$\frac{dE_m}{dt} = \frac{4\pi}{\varepsilon} \int_{x_1}^{x_2} \frac{\partial\rho}{\partial t}\,dx + \frac{4\pi\rho_0}{\varepsilon}\frac{dx_3}{dt}. \tag{21}$$

It can be been from this equation that E_m can have two extrema. Indeed, if $dE_m/dt = 0$, then for $\partial\rho/\partial t > 0$ we have always that $dx_3/dt < 0$, and, conversely, for $\partial\rho/\partial t < 0$, necessarily $dx_3/dt > 0$. Since the derivatives dj/dt and dE_m/dt always have the same sign and reduce to zero simultaneously, we can say the same about j as we can about E_m.

The extrema of E_m and j do not coincide in time either with the extrema of ρ or with the extrema of x_3. All this is characteristic of an oscillatory solution, whence we may conclude that if our system of equations possesses a solution at all, then it is oscillatory in character. A rigorous mathematical proof that the system does, in general, possess a solution is not readily forthcoming. Accordingly, in lieu of a theoretical proof of the existence of a solution, we utilize the experimental fact that periodic variations of current and brightness are indeed observed.

We note that when the voltage dropped across the near-cathode space-charge layer is large, when S_{DI} becomes close to S_I, the density ρ may become negative. in this case, the oscillatory character of the solution of our equations is retained, although the equations themselves, starting from a certain voltage, no longer correspond to the experimental conditions as the assumption of a stationary boundary x_2 is no longer justified. It is precisely under these conditions that ionization domains begin to be formed.

The theory which we have been developing can be checked to some extent by estimating the period of the expected oscillations and comparing it with the value found in experiment.

Let us estimate, first of all, the time during which the space charge is increasing. It is determined by the electron lifetime on deep donors. The probability of electrons being thermally liberated from such donors is small (they have a depth of 1 eV), and can be neglected. In [2] we found the effective cross section for the ionization of donors by electrons emerging from ionization domains; it worked out to be $3 \cdot 10^{-14}$ cm^2. According to our estimates, these electrons have a mean energy much the same as that of the electrons in the present case. In our estimates, therefore, we take S_I to have this value. For a current of 10^{-7} A (oscillations are usually clearly visible at a mean current of this value) and for a crystal of cross-sectional area of 0.04 cm^2, the product $S_I j \approx 0.5$ sec^{-1}.

The de-ionization probability can be calculated in exactly the same way. However, as the threshold energy for donor de-ionization in the investigated crystals is higher by a factor of three, it follows that the effective cross section for donor de-ionization must be several times smaller. For S_{DI} we take the value $5 \cdot 10^{-15}$ cm^2. This gives a de-ionization probability of 0.07. Accordingly, in 1.4 sec, 10% of the donors are de-ionized, i.e., the space-charge density is halved. It follows that the time for the space charge to decay is of the same order as its rise time, and the period of oscillation coincides with the experimental value to within an order of magnitude.

It follows from the above calculations that the space-charge rise and decay times need not by any means be the same. This means that oscillograms of the current and brightness may be saw-tooth in form rather than sinusoidal. As can be seen from Fig. 4a, this is also confirmed by experiment.

The time constants of the processes in wide bandgap semiconductors vary over a wide range: from values much less than a microsecond to values much greater than an hour. The chance of an incorrect theory giving a period of oscillation that differs from the experimentally observed value by less than an order of magnitude is accordingly remote, so that the obtained result can be regarded as a confirmation of the theory.

In this manner, we can assert that the theory of ionization domains has found experimental confirmation not only in the properties of the domains themselves in Mn-activated sodium zincogermanate, but also in the properties of the luminescence observed in these crystals at voltages less than and greater than the voltage at which these domains are clearly visible. The only discrepancy is a quantitative one concerning the relationships between the processes involved in the formation of the ionization domains themselves.

The fact that a wide range of phenomena can be explained from a single point of view testifies very persuasively, of course, to the correctness of the theory. A shadow is cast on the theory, however, by the following interesting observation. All the notions developed in the theory are of a general character, and are in no way connected with the specific structure of the investigated crystals. The numerical values of the parameters involved (effective cross sections, depth of levels, etc.) lie within normal limits. It is not clear, therefore, why such phenomena have not been observed in other wide bandgap semiconductors. We formulated in [2] the conditions necessary for the appearance of ionization domains in semiconductors. These conditions can be satisfied, in particular, in ZnS-Cu, Al crystals with a high impurity content.

A. N. Gorbacheva* kindly prepared for us ZnS-Cu, Al crystals with various Cu and Al concentrations. Application of a dc voltage of a few hundred volts and slight heating (~60°C) resulted in the appearance of a weak luminescence uniformly distributed over the volume of

---

* To whom we are indebted for the crystals used in our investigations.

the crystal, the nature and properties of which are very similar to the background luminescence in sodium zincogermanate crystals.

It would appear, therefore, that our theory is also applicable to other crystals.

LITERATURE CITED

1. M. V. Fok and E. Yu. L'vova, Pis'ma Zh. Éksp. Teor. Fiz., 13:346 (1971).
2. M. V. Fok and E. Yu. L'vova, Tr. FIAN, 68:95 (1973).
3. V. S. Vasilov, V. B. Stopachinskii, and Fan ba Nyan, Kratk. Soobshch. Fiz., No. 4, p. 3 (1972).
4. F. E. Arkhangel'skii, E. Yu. L'vova, and M. V. Fok, Zh. Prikl. Spektrosk., 14:97 (1971).
5. É. V. Devyatykh, E. Yu. L'vova, V. B. Stopachinskii, Fan ba Nyan, and M. V. Fok, Tekh. Poluprovodn., 7:2165 (1973).
6. Fan ba Nyan, Candidate's Dissertation, Physics Institute, Academy of Sciences of the USSR, Moscow (1972).
7. É. V. Devyatykh, E. Yu. L'vova, and M. V. Fok, Fiz. Tekh. Poluprovodn., 8:1813 (1974).
8. É. V. Devyatykh, Candidate's Dissertation, Physics Institute, Academy of Sciences of the USSR, Moscow (1974).
9. V. L. Levshin, É. Ya. Arapova, A. I. Blazhevich, et al., Tr. FIAN, 23:64 (1963).

DOMAIN LUMINESCENCE OF ZINC SULFIDE

A. N. Georgobiani and P. A. Todua

Homogeneous single crystals of ZnS are found to exhibit low-inertia low-voltage prebreakdown electroluminescence (including UV electroluminescence) at voltages corresponding to a mean field in the
crystal of 10^3 V/cm. A negative differential conductivity effect accompanied by acoustoelectrical
instability and domain formation is observed. The electroluminescence and the instability are shown
to be related. The electroluminescence is due to the concentration of the electric field in the domain
up to a value sufficient to excite electroluminescence by electrical breakdown processes.

INTRODUCTION

There is great interest at the present time in semiconducting materials with a wide
forbidden band. Such materials find important applications in the construction of active and
passive circuit elements capable of operating at high temperatures over a wide range of voltages and, especially, in the construction of sources and receivers of short-wavelength light.
There is particular interest in sources in which electrical energy is converted directly into
light as a result of luminescence.

As is well known, two conditions must be met in a semiconductor that is to generate
short-wavelength radiation: the forbidden band must be sufficiently wide, and the probability
of electronic radiative transitions must be sufficiently high. In this respect there is interest
in II-IV semiconductors, particularly zinc sulfide, of which the energy gap (3.7 eV [1]) corresponds to ultraviolet light and in which the probability of emission of a light quantum is large,
both for interband transitions (insofar as they are direct) and also for transitions involving
impurities (the photoluminescence yield of the best luminophors based on ZnS is close to 100%).

Many papers have been devoted to the study of zinc sulfide. The high energy efficiency
of ZnS luminescence finds wide application in industry. Zinc sulfide-based luminophors have
been produced which are currently being used in illuminated display panels and screens, light
amplifiers, optrons, and many other electroluminescent devices. Zinc sulfide is the basis of
luminophors for television screens and oscilloscopes. In the form of thin layers it is used to
make light filters and selective optical mirrors. It is employed in nuclear particle counters
and in efficient up-converters of long-wavelength radiation, whereby infrared and even radio-
frequency radiation can be made visible.

Most papers published to date on the photo-, cathodo-, and electroluminescence of zinc
sulfide have been concerned with polycrystalline samples. Comparatively little work has been
done on single-crystal ZnS, although the number of papers appearing on this topic, and so also
the interest in ZnS single crystals, is steadily growing.

Recombination luminescence of homogeneous ZnS single crystals accompanying electrical breakdown is reported in [2]; p-type conductivity in ZnS is obtained in [3, 4]; and an electroluminescing p−n junction is reported in [3, 5].

At the present time work is being carried out with a view to improving the manufacture of both pure and perfect ZnS single crystals and ZnS single crystals doped with various donor and acceptor impurities; techniques are being developed for producing solid solutions of ZnS with other II-IV compounds, solutions such as ZnS_xSe_{1-x}, $Zn_xCd_{1-x}S$, and so on [6], including those with a bandgap gradient promising for optoelectronics. Materials of this sort have been used as a basis for metal−dielectric−semiconductor, metal−oxide−semiconductor, metal−insulator−semiconductor and other structures, p−n homo- and heterojunctions, and so on, all of which can find many applications in diverse branches of science and industry besides optoelectronics − for example, as standard sources, receivers, and converters of radiation in metrology and measuring techniques, as sensing elements of setups for measuring high electrical voltages, and so on.

Zinc sulfide, both pure and compensated, has a resistivity that is quite high, corresponding rather to that of a good dielectric [2]. However, in the manufacture of a whole series of semiconductor devices, it is starting materials with a high conductivity that are required. In this connection we have made a study of zinc sulfide single crystals doped with various donor impurities. In ZnS samples with a comparatively low resistivity we discovered a low-voltage electroluminescence of prebreakdown type, induced by electrical instability [7]. In the present paper we report a study of the electrical characteristics of such crystals and the properties of their electroluminescence.

CHAPTER I

PROCESSES RESPONSIBLE FOR EXCITATION OF ELECTROLUMINESCENCE IN SEMICONDUCTORS

Two types of electroluminescence can be distinguished depending on the processes whereby it is excited: injection electroluminescence, and prebreakdown electroluminescence. In the first case, minority carriers are injected into the semiconductor, and the luminescence is the result of their recombination with majority carriers, both directly and via luminescence centers. An example of this sort of electroluminescence is the emission of a forward-biased p−n junction (Fig. 1). Injection electroluminescence can usually be excited at relatively small voltages, corresponding to the energy width of the forbidden band.

In prebreakdown electroluminescence, on the other hand, electron−hole pairs are produced and luminescence centers are excited as a result of the processes causing the electrical breakdown of the semiconductor (field ionization, impact ionization). Processes of this sort can occur in most electroluminors and also in reverse-biased p−n junctions. The field intensities required before such processes can develop, and so also before prebreakdown electroluminescence can be excited, are of the order of 10^5-10^7 V/cm [8-11].

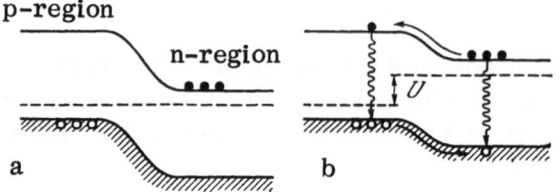

Fig. 1. Energy level diagram of electroluminescing p−n junction. (a) With no external voltage; (b) with forward bias. The wavy arrow denotes direct recombination of an electron and a hole.

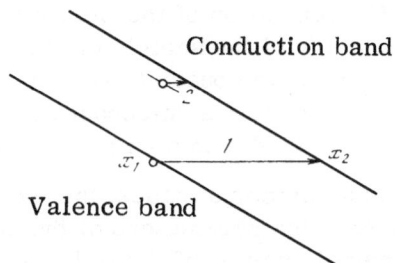

Fig. 2. Schematic representation of field-ionization processes (Zener breakdown). 1 and 2
are tunnel transitions associated with the ionization of the crystal lattice itself and of defects
respectively.

Injection sources of light based on the following II-IV compounds have been reported:
ZnTe [12], ZnSe [13-15], ZnS [3, 5], where bands corresponding to interband transitions were
observed in the emission spectra of ZnSe and ZnS diodes.

Nothing has been reported in the literature on the construction of injection lasers based
on II-IV compounds. The reason for this is the difficulty of obtaining a large excitation level.
For electron-beam excitation, laser action has been observed in most of these compounds:
CdS [16], CdTe [17], CdSe [18], ZnS [19], ZnSe [20], CdS_xSe_{1-x} [21], $Zn_xCd_{1-x}S$, $Zn_xCd_{1-x}Se$ [22].

The main processes responsible for the electrical breakdown of a semiconductor and
capable of giving rise to prebreakdown electroluminescence are field and impact ionization.

In thie first case, nonequilibrium current carriers are produced as a result of the ionization of the crystal lattice and localized centers via electron tunneling through the potential
barrier into the conduction band. Tunneling becomes possible due to the slope of the energy
bands in a strong electric field (Fig. 2). Electric fields of the order of 10^6-10^7 V/cm are required in order to realize this effect [10].

A strong electric field can also induce electroluminescence through impact ionization of
the crystal lattice and luminescence centers (Fig. 3). Calculations show that fields of the order
of 10^5-10^6 V/cm are required before such processes can get under way [8, 9, 11]. Accordingly,
unless there are special conditions (e.g., field applied to a very thin layer, when it is difficult
for an avalanche to develop), impact ionization occurs at much smaller fields than tunnel ionization.

Usually, however, prebreakdown electroluminescence of single crystals and powdered
luminophors based on ZnS is observed at voltages corresponding to mean fields much smaller
than this, namely $\sim 10^3$-10^5 V/cm. It has been shown that the electric field is concentrated in
individual parts of the crystal, where it reaches an effective value of $\sim 10^6$ V/cm [23]; the
luminescence is nonuniformly distributed over the volume, as the region of increased field
intensity glows more brightly than the main bulk of the crystal by several orders of magnitude.

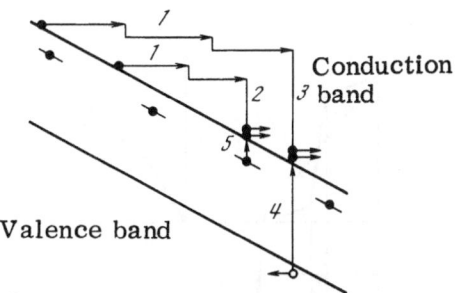

Fig. 3. Schematic representation of processes
of electron acceleration and impact ionization
(avalanche breakdown). 1 shows an electron
being accelerated (the steps denote loss of energy as a result of scattering events); 2, 3 correspond to loss of energy by electrons as a
result of impact ionization of the crystal lattice
(4) and a defect (5).

Concentration of the field may come about for several reasons; it may be connected with the geometry of the sample or the electrodes, with nonuniformity of the electrical conductivity or the dielectric constant, with the presence in the sample of potential barriers, e.g., contact barriers of metal−semiconductor or semiconductor−semiconductor type, and potential barriers associated with p−n homo- and heterojunctions.

In a number of cases, the electric field in the crystal can be concentrated as a result of a change of the parameters of the current carriers brought about by the field itself. Starting from certain values of the field, the field distribution in the sample becomes very nonuniform, Ohm's law ceases to be obeyed, and the current in the circuit containing the sample starts to oscillate. This phenomenon has been termed "electrical instability."

Electrical instability in semiconductors is receiving considerable attention at the present time (see [24], for example) in connection with various possible practical applications (Gunn oscillators, and so on). Let us consider this matter in more detail.

In an isotropic spatially homogeneous semiconductor with unipolar conductivity, the current density in the one-dimensional case can be written in the form

$$j = en\mu E, \tag{1}$$

where j is the current density, e is the mobile carrier charge, n and μ are the concentration and mobility of the current carriers, and E is the electric field intensity.

The differential conductivity of the semiconductor, σ_i, can be expressed in the form

$$\sigma_i = \frac{dj}{dE} = en\mu \left(1 + \frac{d \ln \mu}{d \ln E} + \frac{d \ln n}{d \ln E} \right). \tag{2}$$

It can be seen from this expression that a deviation from Ohm's law can occur if the mobility of the current carriers or their concentration (or both) depends on the field intensity. For certain values of E the functions $\mu(E)$ and (or) $n(E)$ may be such that σ_i becomes negative, when the volt−ampere characteristic will possess a falling region (an N-type volt−ampere characteristic) [25, 26]. A spatially homogeneous sample may then become electrically unstable and divide into regions in which the electric field is strong or weak (Fig. 4). In this case, the oscillations of current that appear in the circuit containing the sample can be ascribed to the periodic formation and displacement within the crystal of mobile electric domains [24], regions of high resistance in which the field in strong. Domains of this sort have been observed in a number of crystals: GaAs [27, 28], InP [27], CdS [29], CdTe [14], Ge [30].

These domains move in the same direction as the majority current carriers. The shape of the domain usually remains unchanged during its motion within the crystal, and its velocity of motion remains almost constant. If the characteristic times of nucleation and decay of a domain are very much less than its transit time across the crystal, then the period of the

Fig. 4. Nonuniform distribution of electric field within sample. (a) Voltage distribution along sample (X is region of enhanced electric field intensity); (b) typical shape of N-type volt−ampere characteristic.

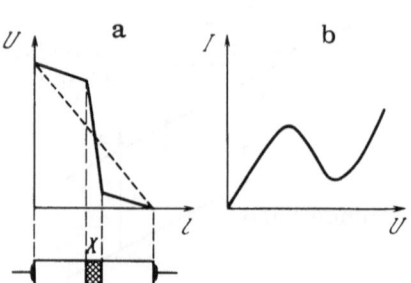

current oscillations in the circuit will be determined by the transit time alone and will equal the length of the sample divided by the velocity of motion of the domain [24].

There are a number of electrical instabilities that can give rise to an N-shaped current-voltage characteristic. Firstly, there is the Gunn effect [27]. In this case, the appearance of electrical instability in an n-type semiconductor is connected with a decrease of the electron mobility which occurs when the electrons are "heated up" in sufficiently strong electric fields, when they pass from the absolute minimum of the conduction band (where they have a smaller effective mass) to a secondary minimum (greater effective mass).

In n-type cadmium telluride, the Gunn effect is observed at fields of the order of 10^4 V/cm, the frequency of oscillation of the current being of the order of gigahertz and depending on the length of the sample [31].

Acoustoelectric instability, involving the generation of acoustic phonons, has been observed in piezoelectric semiconductors when the drift velocity of conduction electrons exceeds the velocity of sound [32-34]. This instability is also connected with the field dependence of the mobility. The field intensities at which this effect has been observed in a number of semiconductors, including II-IV compounds, lie in the range 10^3-10^4 V/cm.

An instability known as recombination instability has been observed in a number of crystals. This instability is connected with a decrease in the free-carrier concentration that occurs in sufficiently strong electric fields, due to dependence of the capture coefficients for hot electrons on the electric field. This is possible, in particular, for capture by a center of like charge, e.g., capture of an electron by a negatively charged center [35-36].

A temperature-electric instability has been observed in CdSe and some other II-IV compounds. In this case, the concentration of nonequilibrium carriers produced by illumination of the sample is decreased due to the rise of temperature of the crystal as it heats up in the electric field (temperature quenching of the photoconductivity) [37].

The formation of electric domains and the resultant concentration of the electric field probably explains why the electroluminescence of CdS without contact barriers extends to photon energies close to the width of the forbidden band for voltages corresponding to a mean field in the crystal of only $\sim 4 \cdot 10^2$ V/cm [38]. Calculations show that, if there were no such domains, electric fields greater than this by three orders of magnitude would be required.

CHAPTER II

EXPERIMENTAL

1. Sample Preparation

We investigated ZnS-I single crystals grown in the Physics Institute of the Academy of Sciences by A. V. Lavrov by the iodide transport method at T = 700-800°C. The starting material was powdered zinc sulfide of luminophor quality, which was first purified from chlorine and sulfate impurities by annealing in an atmosphere of dry ammonia at T = 900°C. To increase the electrical conductivity, the crystals were treated for 24-48 h in a zinc melt at T = 950°C. After annealing, the crystals were etched in pyrophosphoric acid at 180-200°C for 0.5-2 min, followed by washing in distilled water and alcohol. The resistivity of the samples before annealing was 10^8-10^9 Ω · cm at T = 300°K; after annealing it was 10^3-10^4 Ω · cm, which is much less than the resistivity of ZnS crystals that is usually observed [2]. Uniformity of resistance of the samples was checked by splitting off wavers and etching. For the experiments we selected plane-parallel platelets of dimensions $\sim 3 \times 2$ mm and thickness ~ 1 mm.

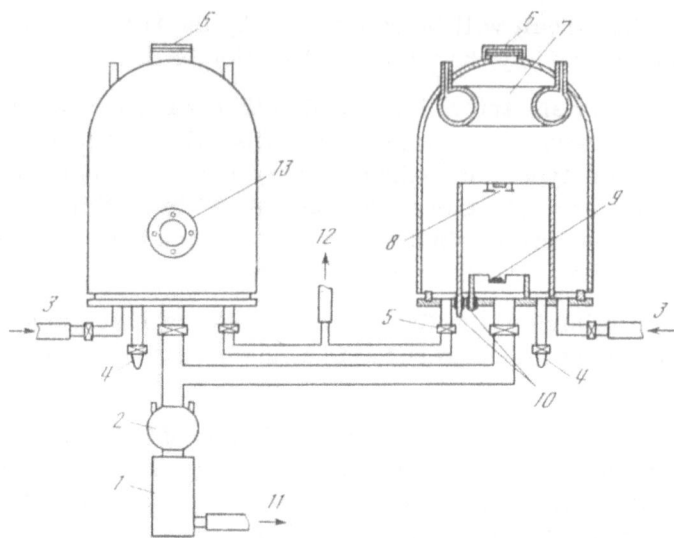

Fig. 5. Schematic diagram of vacuum deposition apparatus. (1) DOV-500 diffusion pump; (2) nitrogen trap; (3) needle valve for introduction of inert gas; (4) air inlet tap; (5) tap for evacuation by auxiliary forevacuum pump; (6) observation window; (7) toroidal nitrogen trap; (8) crystal in mount on heater; (9) indium in tantalum boat; (10) vacuum-tight electrical leads; (11) to main forevacuum pump; (12) to auxiliary forevacuum pump; (13) observation window.

Measurement of the sign of the thermo-emf showed that the conductivity is n-type. Ohmic contacts were made using spectrally pure indium. The contacts were deposited in a vacuum deposition apparatus (Figs. 5 and 6) in a vacuum of not less than 10^{-6} mm Hg. The apparatus provided for decontamination of the surface of the samples in an inert-gas glow discharge, deposition of the metal, and its subsequent infusion in a single technological cycle. Best results were obtained for ion etching times of $\approx$40 min, indium deposition at 20-30 Å/min (15-20 min), and infusion of the contact at 600°C for 10 min. Later we discovered that a satisfactory ohmic contact can be obtained by infusing the indium in a nitrogen atmosphere.

Electrical forming by means of a short pulse of current (capacitor discharge) completed the fabrication of the contact. The initial part of the volt-ampere characteristic of a ZnS-I single crystal treated in a zinc melt (Fig. 7) confirms the excellent quality of the ohmic contacts made in the above manner.

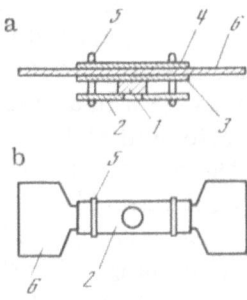

Fig. 6. Heater and mount for crystals. (a) View from front; (b) view from below. (1) Sample; (2) mica masks; (3, 4) mica insulating spacers; (5) spring clips; (6) tantalum heater.

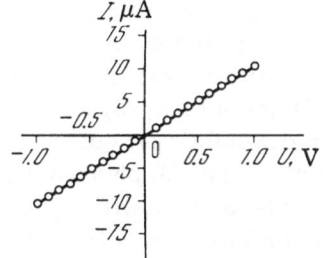

Fig. 7. Initial part of volt−ampere characteristic of ZnS-I crystal treated in zinc melt.

2. Investigation of Electroluminescence and Electrical Characteristics of ZnS Single Crystals

The samples were investigated both at room temperature and at liquid nitrogen temperature. For this purpose we constructed a special cryostat (Fig. 8). It comprises a quartz Dewar
provided with two quartz optical windows viewing at right angles. The crystal, fixed to the
holder by means of indium solder, is located directly in the liquid nitrogen. The Dewar is
placed in a metal holder mounted on a rider which can be moved along a tetrahedral rail. The
holder is provided with a manipulator, by means of which it can be moved smoothy in three
mutually perpendicular directions and rotated around its axis. This facilitates the adjustment
of the optical system.

The spectral investigations were performed on a photometric setup (Fig. 9) based on
a ZMR-3 mirror monochromator with a quartz prism (wavelength range 220-2500 nm) and
photomultipliers for various spectral ranges: FÉU-39 (spectral sensitivity range 160-600 nm);
FÉU-14A (300-750 nm); FÉU-79 (300-830 nm); FÉU-28 (400-1100 nm); FÉU-30 (300-600 nm).
The latter photomultiplier is used to study fast processes, of duration down to 10 nsec. The
spectrum is scanned automatically. The setup was calibrated using a standard photometric
lamp with a color temperature $T_{col}= 3000°K$.

Electroluminescence of ZnS-I single crystals treated in a zinc melt is observed both
under dc and pulsed excitation conditions. However, to prevent the samples from heating up,
electroluminescence was excited by square pulses of variable amplitude of duration 3 μsec
and off-duty ratio 10^3. Under such conditions electrical breakdown will occur before thermal

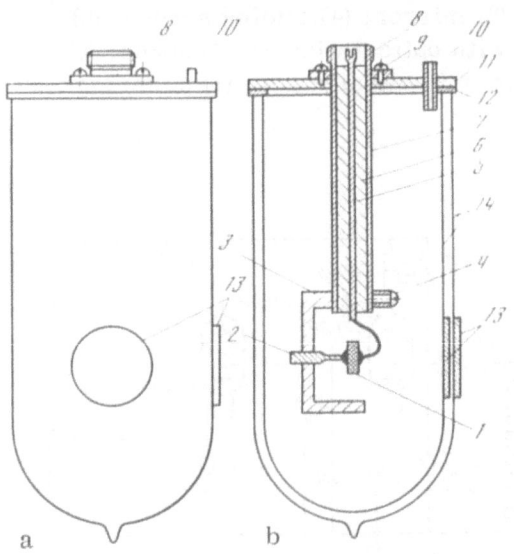

Fig. 8. Schematic diagram of nitrogen cryostat
used to investigate ZnS single crystals. (a)
Overall view; (b) cross-sectional view. (1) Crystal with ohmic contacts; (2) crystal holder; (3)
supporting bracket; (4) fastening screw; (5)
copper conductor; (6) Teflon insulation; (7)
copper tube; (8) rf socket; (9) fixing screw; (10)
tube for filling with liquid nitrogen; (11) metal
cap; (12) rubber gasket; (13) quartz windows;
(14) walls of quartz Dewar vessel.

Fig. 9. Block diagram of photometric setup for
investigating electroluminescence spectra. (1)
Generator (type G5-15); (2) generator of square
voltage pulses; (3) sample under investigation
(in cryostat); (4) quartz lens; (5) ZMR-3 monochromator; (6) photomultiplier; (7) USh-10
broad-band amplifier; (8) synchronous detector;
(9) PSI-02 automatic recording potentiometer.

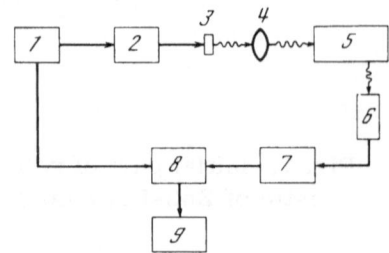

breakdown, even in ionic crystals [39]. Also, the reduced heating of the crystals means that
they can be placed directly in the liquid nitrogen. The fact that the contacts are ohmic means
that there is no heating in the vicinity of the contacts.

The narrow slit of the monochromator had to be used when making detailed studies of
the spectra; this, along with the large off-duty ratio, led to a considerable degradation of the
signal/noise ratio. The signal/noise ratio was improved by connecting a synchronous detector
between the amplifier and the recording device (see Fig. 9).

The voltage−brightness characteristic (i.e., the dependence of the integrated electroluminescent brightness on applied voltage) at low brightness levels was measured using a
photon counting setup (Fig. 10) by means of which a few quanta of radiation per second could
be recorded. The dark current was reduced by placing a specially selected FÉU-14B photomultiplier in the Dewar with the liquid nitrogen. The FÉU-14B output was connected to a pulse
amplifier designed around 6Zh9P tubes (Fig. 11). The recording instrument was a PS-10000
counter. An M-95 microammeter was used for large luminescence intensities.

The time dependence of the electroluminescence was studied by simultaneously oscillo-
graphing the exciting voltage pulse and the electroluminescent brightness of the investigated

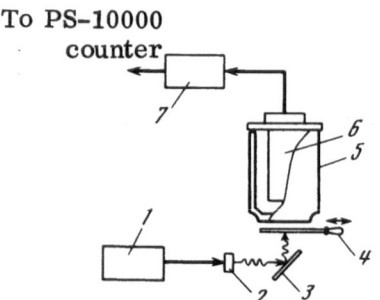

Fig. 10. Block diagram of setup for investigating voltage−brightness characteristics. (1)
Generator of square voltage pulses; (2) sample
(in cryostat); (3) mirror; (4) sliding screen; (5)
Dewar vessel with optical window in bottom; (6)
photomultiplier in liquid nitrogen; (7) pulse amplifier.

Fig. 11. Circuit diagram of pulse amplifier.

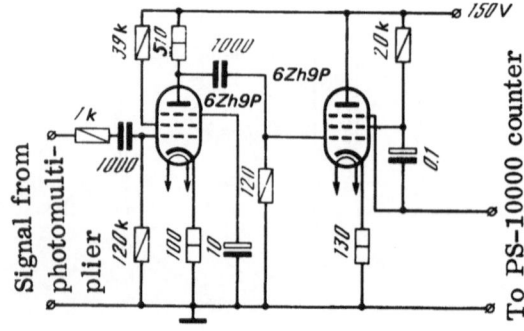

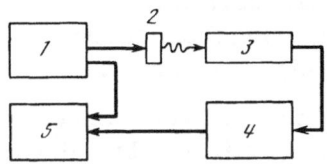

Fig. 12. Block diagram of setup for investigating time characteristics of electroluminescent emission. (1) Generator of square voltage pulses; (2) sample (in cryostat); (3) photomultiplier; (4) broad-band amplifier; (5) double-beam oscilloscope.

single crystals (Fig. 12). The time resolution of the setup was checked using silicon carbide light-emitting diodes, the inertia of which was less than 10^{-8} sec.

A block diagram of the setup used to investigate the space−time characteristics of the electroluminescence is shown in Fig. 13. The sample is connected in series with a reference resistor R_1, the voltage drop across which (proportional to the current pulse) is fed to one of the inputs of a double-beam oscilloscope. For the current in the circuit crystal/reference resistor to be determined solely by the crystal resistance, we must have, of course, $R_1 \ll R_0$. In the image plane of the luminescing crystal there is located a rectangular screen that can be displaced along the image by means of a micrometer transmission. By fixing the screen in a definite position, we can feed to the photomultiplier input the emission from a definite part of the luminescing crystal. The simultaneous display on the oscilloscope screen of the current

Fig. 13. Block diagram of setup for investigating space−time characteristics of electroluminescent emission. (1) Generator of square voltage pulses; (2) quartz lens; (3) screen, displaceable by micrometer transmission in image plane AA'; (4) photomultiplier; (5) broad-band amplifier; (6) double-beam oscilloscope; R_0, crystal under investigation (in cryostat); R_1, reference resistor.

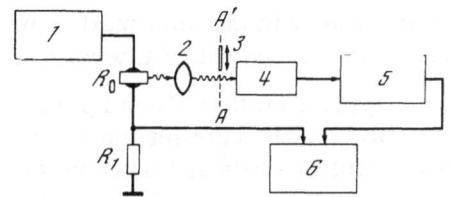

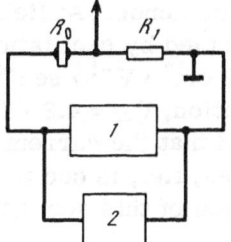

Fig. 14. Block diagram of setup for measuring static volt−ampere characteristic. (1) Generator of square voltage pulses; (2) V4-1A pulse millivoltmeter with 1:100 divider; R_0, crystal under investigation (in cryostat); R_1, reference resistor; the arrowed lead goes to a pulse millivoltmeter (type V4-3) or an oscilloscope.

Fig. 15. Block diagram of setup for measuring pulsed volt−ampere characteristics. (1) Generator; (2) S1-16 oscilloscope; R_0, investigated crystal; R_1, reference resistor.

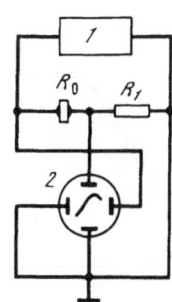

pulse and the light pulse from the whole sample or from a certain part of it yields information on the phase or time relationships between the luminescent emission from an individual part of the sample (or all of it) and the current in the circuit containing the sample.

Block diagrams of the setups for measuring the static and pulsed volt−ampere characteristics of ZnS single crystals are shown in Figs. 14 and 15 respectively. In both cases, the voltage drop across reference resistor R_1 is used as a measure of the current through the sample.

CHAPTER III

ELECTRICAL AND ELECTROLUMINESCENT PROPERTIES OF ZnS SINGLE CRYSTALS

1. Volt − Ampere Characteristics

Zinc sulfide crystals grown by a variety of methods usually possess n–type conductivity. Nonetheless, with a view to obtaining general information on our samples, both those treated in a zinc melt and those that were not subjected to this sort of heat treatment, we measured the sign of the thermo-emf. As was expected, n-type conductivity was observed in all cases, testifying that doping with iodine during growth and with zinc during the heat treatment, which leads to increased electrical conductivity, corresponds to doping the semiconductor with a donor impurity. The reduction of the resistivity of ZnS–I by four to five orders of magnitude after treatment in the zinc melt is probably due to the filling of zinc vacancies and also to an increase in the content of interstitial zinc, which is a donor.

A typical static volt−ampere characteristic is shown in Fig. 16. It can be seen that at small voltages the current increases in accordance with Ohm's law (ohmic part of volt−ampere characteristic) while at larger voltages, corresponding to a mean field in the crystal of $\gtrsim 10^3$ V/cm (crystal thickness ≈ 1 mm), the current saturates.

Non-ohmic saturation of the current in low resistivity ($\rho \sim 10\ \Omega \cdot$ cm) CdS crystals at a critical field $E_{cr} \approx 1.6 \cdot 10^3$ V/cm was first reported by Smith [40], who established that this effect originates in the body of the crystal and is not a contact phenomenon. At fields equal to or greater than E_{cr}, the observed saturation effect was accompanied by oscillations of current. Remembering that the drift mobility in these samples is 300 cm$^2 \cdot$ V$^{-1} \cdot$ sec^{-1}, it turns out that the critical drift velocity of the carriers at current saturation, $V_{cr} = 4.8 \cdot 10^5$ cm/sec, close to the velocity of sound in cadmium sulfide. Smith suggested that the current saturation effect is a direct consequence of the amplification of acoustic waves, i.e., is due to transfer of energy from the electrons to a traveling phonon wave. Saturation of this sort has also been

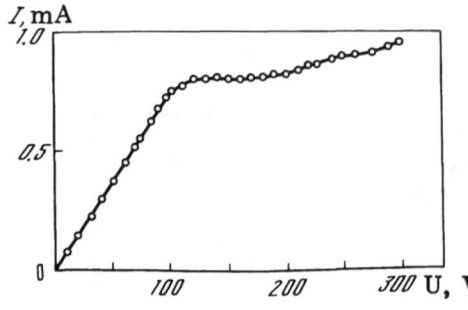

Fig. 16. Static volt−ampere characteristic of a ZnS sample.

observed in CdS and CdSe crystals of higher resistivity [41]. According to Moore's results [42], $V_{cr} = 4.4 \cdot 10^5$ cm/sec. Current saturation has also been observed in ZnS [43]; this saturation is not, however, connected with the acoustoelectric effect, as the accompanying oscillations of current are characterized by low frequencies. The nature of the slow domains responsible for this instability was not determined.

In our ZnS-I crystals, the critical field $E_{cr} = (1.2 \pm 0.1) \cdot 10^3$ V/cm. Remembering that the electron drift mobility in ZnS at liquid nitrogen temperature is 300 cm^2/V · sec [44], the critical velocity V_{cr} works out to be $(3.6 \pm 0.3) \cdot 10^5$ cm/sec, which is in good agreement with the velocity of sound in ZnS ($3.4 \cdot 10^5$ cm/sec [45, 46]).

The acoustic waves and the electron distribution can be regarded as two interacting systems between which energy and momentum exchange occur. In the case of interaction of the electrons with a single acoustic mode of angular frequency ω, the absorption coefficient of the acoustic oscillations per unit path α is found to be [41]

$$\alpha = \frac{C^2 m^{*2} \omega}{2\pi \hbar^3 \rho V_s^2} y \left(1 - \frac{V_d}{V_s}\right), \tag{3}$$

where ρ is the density of the crystal, C is the electron−phonon coupling constant, V_s is the acoustic-wave velocity, V_d is the electron drift velocity, $y = \frac{1}{2} n_0 \hbar^3 / (2\pi m^* kT)^{3/2}$ is the "degeneracy parameter" (equal to unity if the electron distribution is degenerate, and many times greater than unity if Maxwell−Boltzmann statistics apply), and n_0 is the equilibrium electron concentration. According to this expression, the absorption α depends linearly on the drift velocity and becomes negative if $V_d > V_s$, when phonon emission begins to prevail over phonon absorption, i.e., acoustic amplification occurs.

Hutson et al. [32] noted that the acoustic energy in CdS under these conditions increased. Further investigations showed that current saturation in CdS is accompanied by the spontaneous generation of acoustic waves [47, 48]. The discovery of a correlation between current saturation and an increase of acoustic energy is strong evidence in favor of the acoustoelectric effect.

An oscillogram of the pulsed volt−ampere characteristic of a zinc-melt-treated ZnS-I single crystal is shown in Fig. 17. It can be seen that the characteristic has a falling portion. It begins at a voltage corresponding to a mean field in the crystal of ~10^3 V/cm.

For electrical instability originating in the bulk of the sample, the pulsed volt−ampere characteristics have a falling portion or a portion of current saturation, while the static volt−ampere characteristics have a portion of current saturation [24-26]. The appearance of volt−ampere characteristics of this sort may be an indication of electrical instability, manifesting

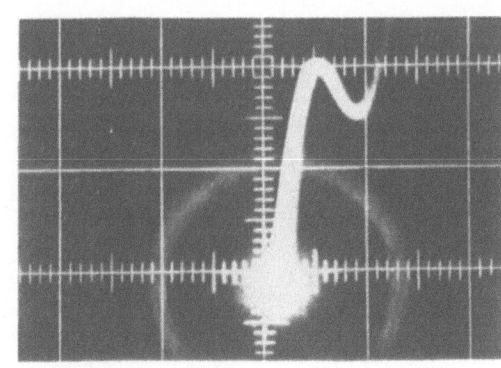

Fig. 17. Pulsed volt−ampere characteristic.
Scale: horizontal 30 V/div; vertical 35 μA/div.

itself externally in a number of cases in oscillations of the current through the crystal at a constant voltage across it.

The nucleation and motion of an acoustoelectric domain can be envisaged in the following manner. When $V_d > V_s$ the electrons give up some of their energy to phonons and their mobility is accordingly reduced. The resistance thereupon increases, since $\rho = 1/ne\mu$, the increase being greater in that part of the crystal where, for some reason or another, the field is greater (e.g., near an electrode or macroscopic nonuniformity in the bulk of the crystal), which causes the electric field intensity in this part to become greater still. As a result of this increase in the field intensity, a large proportion of electrons can be accelerated to velocities greater than V_s; this eventually results in the intensity in the strong-field region (domain) increasing to such an extent that electrical breakdown processes are able to develop. This leads, firstly, to luminescence, and secondly, to the electrical conductivity of this part of the crystal increasing to such an extent that no further concentration of electric field there is possible. The field then either becomes uniformly distributed over the volume of the crystal (decay of domain) or is concentrated in another part of the crystal (movement of domain).

2. Electroluminescence

The investigated samples had a bright blue photoluminescence [49, 50]. Electroluminescence was excited using voltage pulses of duration 3 μsec and off-duty ratio 10^3; the pulse amplitude could be varied from 0 to 300 V, which, for a sample thickness of ≈ 1 mm, corresponds to a mean field in the crystal of 0-3 · 10^3 V/cm.

The samples began to luminesce at voltages corresponding to mean fields in the crystal of $\sim 10^2$ V/cm, which is very much less than the fields ordinarily required for electroluminescence of ZnS. To check that this emission is not due to electrical microbreakdowns of gas in any thin cracks or bubbles that may be formed during the growth and heat treatment of the crystals or the preparation of the contacts, we photographed the electroluminescence spectrum using a quartz spectrograph. The spectrum was found to be devoid of the narrow lines characteristic of an electrical discharge in a gas.

The electroluminescence spectrum of single crystals (Fig. 18) was recorded for mean fields in the crystal of $\sim 10^3$ V/cm. At room temperature this spectrum has the form of a broad band with a weakly expressed maximum at 465 nm. The intensity of the emission increases by a factor of more than 30 at liquid nitrogen temperature, and, besides the blue band, there also appears a red band peaking at around 700 nm and an ultraviolet band extending down to 330 nm, which corresponds to the intrinsic absorption edge. It is interesting that this electroluminescence is observed at fields two to three orders of magnitude less than are required to impact-ionize the crystal lattice and luminescence centers.

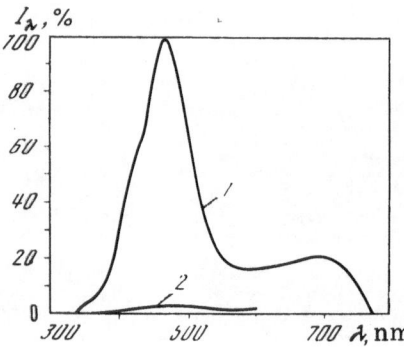

Fig. 18. Electroluminescence spectrum. (1) T = 77°K; (2) T = 300°K.

Microscope studies (resolution not less than 0.5 μm) of the electroluminescing crystals did not reveal any nonuniformity in the distribution of the glow over the volume, from which we may conclude that there are no stationary inhomogeneities present which can lead to concentration of the electric field. The fact that the contacts are ohmic probably excludes the possibility of minority carrier injection and luminescence due to their recombination with majority carriers, either directly or via luminescence centers. Nonetheless, the presence in the electroluminescence spectrum of light of wavelength λ = 330 nm, which is close to the energy gap of ZnS, seems to indicate that the emission is due to the recombination of electrons with holes injected from the contacts. In this event, one would expect the luminescence to be localized near the anode part of the crystal (since holes in ZnS are not very mobile), the size of the luminescing region being comparable with the diffusion length or the drift length of the holes injected from the anode (depending on which of these lengths is the greater).

The quantities L_{diff} and L_{dr} can be estimated from

$$L_{diff} = (D_h\tau_h)^{1/2},$$ (4)

$$L_{dr} = E\mu_h\tau_h,$$ (5)

where D_h, μ_h, and τ_h are respectively the hole diffusion coefficient, mobility, and lifetime.

The hole mobility μ_h measured [51] at T = 300°K is 5 cm^2/V · sec. Unfortunately, no data is available on the hole mobility in ZnS at T = 77°K, although in order of magnitude it is unlikely to exceed 10 cm^2/V · sec; we then have that $D_h = \mu_h kT/e \sim 6 \cdot 10^{-2}$ cm^2/sec. A reasonable value of the hole lifetime is not more than 10^{-7} sec [52], so that we obtain $L_{diff} \sim 10^{-14}$ cm and $L_{dr} \sim 10^{-3}$ cm at a field intensity of $\sim 10^3$ V/cm (it is at such intensities that current saturation and electroluminescence are observed [53, 54]). As the glow in the experiments is not localized at a distance of a few drift lengths ($L_{dr} > L_{diff}$) from the anode, it follows that it is not the result of recombination of electrons with injected holes.

Information on the mechanism responsible for the electroluminescence can be deduced from the form of the voltage–brightness characteristic B(U) (i.e., the dependence of the integrated brightness of the electroluminescence on applied voltage). We investigated this dependence over a wide range of brightnesses with the aid of a method in which photoelectrons were counted using a photomultiplier, thereby enabling weak light fluxes to be recorded. The results of the measurements are presented in Fig. 19. It can be seen that this dependence, when plotted in coordinates log B and $U^{-1/2}$, is effectively a straight line; it can be expressed as

$$B = B_0 \exp\left(-b/\sqrt{U}\right),$$ (6)

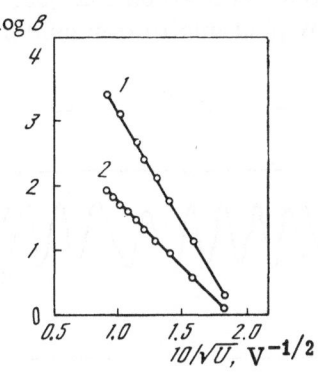

Fig. 19. Dependence of integrated electroluminescent brightness on applied voltage. (1) Liquid-nitrogen temperature; (2) room temperature.

where B is the brightness, and B_0 and b are quantities that are independent of voltage.

A B(U) dependence of this sort is usually observed in the prebreakdown electroluminescence of a variety of ZnS-based samples, including electroluminescent capacitors [55, 56]; inhomogeneous crystals [57]; p—n junctions [3]; ZnS-Cu_2S heterojunctions [3]. It is usually assumed that in these cases the field is concentrated in Mott—Schottky barriers and p—n junctions. For a constant space-charge density in the barrier, the field is connected with the voltage by the relationship $E \sim \sqrt{U}$. Utilizing this relationship, we can write (6) in the form

$$B \sim \exp\left(-b_1/E\right). \tag{7}$$

This dependence is usually associated with electrical breakdown processes, as the probability of impact- and field-ionization processes is also characterized by a field dependence of this sort.

The discovery of current oscillations in the crystal circuit and correlated oscillations of electroluminescent brightness is of special interest. The observed oscillations of current were consistent with electrical instability induced by the appearance, motion, and decay of high-field electric domains. It should be noted that current and brightness oscillations occurred at electric field intensities at which the volt—ampere characteristic deviated from Ohm's law. A correlation of this sort indicates the important role played by domains in the excitation of electroluminescence in our crystals.

Unfortunately, how the space-charge density is distributed within the domains is unknown. If we assume that the space charge within the domain is uniformly distributed over the width of the domain, it can be shown that $E \sim \sqrt{U}$. The field dependence of the electroluminescent brightness in our case will then also have the form (7). In one way or another, the dependence B(U) turned out to be typical for prebreakdown electroluminescence, which indicates that the field in the domain becomes sufficiently concentration for prebreakdown electroluminescence processes to develop.

3. Time Characteristics of Electroluminescence
of ZnS Single Crystals

Further information on the nature of the observed recombination radiation was obtained from a study of the electroluminescence kinetics. Oscillograms of the electroluminescent brightness and the exciting voltage taken during a single sweep are shown in Fig. 20. The duration of the leading and trailing edges of the exciting voltage pulse is 0.2 μsec. It can be seen that the edges of the brightness oscillogram are effectively undelayed relative to the exciting voltage pulse, indicating that the relaxation time of this electroluminescence τ_{relax} is ≤ 0.2 μsec. This is two to three orders of magnitude less than the relaxation time of powdered electroluminophors based on ZnS [55, 56]. The oscillations on the brightness oscillogram are not induced by photomultiplier noise.

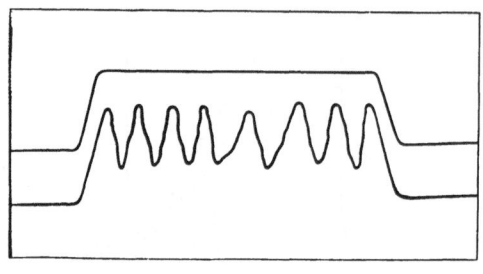

Fig. 20. Oscillograms of electroluminescent brightness (bottom) and exciting voltage (top). The oscillograms were taken simultaneously in a single sweep. Parameters of exciting voltage: amplitude 120 V, duration 3 μsec.

We have suggested that electrical instability is responsible for the excitation of the ob- served electroluminescence; accordingly, the time characteristics of the electroluminescence and the current oscillations must be investigated.

The nature of the electrical instability was assessed from the waveform of the current oscillations in the circuit containing the electroluminescing sample, and the connection between the instability and low-voltage electroluminescence was made on the basis of a study of the current oscillations along with the brightness oscillations. Clearly, in pulse measurements, the duration of the exciting voltage pulse must be much greater than the duration of the oscilla- tions, so that the electric domain has time to nucleate, move along the crystal, and decay during the time for which the exciting voltage pulse is applied.

A strong correlation is observed between the current oscillations through the crystal (at a constant amplitude of the voltage pulse across the contacts) and the electroluminescent brightness oscillations (Fig. 21). A similar effect was observed in [58] in the electrolumines- cence of CdS single crystals, the electroluminscence spectrum of which contains radiation peaking at $\lambda = 5625$ Å, which corresponds approximately to the width of the forbidden band of CdS at T = 77°K (the well-known green emission of cadmium sulfide). As mentioned in § 2 of this chapter, emission close to the energy gap of ZnS was observed in our case as well.

The time taken by a domain to traverse the length of the crystal was determined from the duration of the oscillations ($\tau = 0.31 \pm 0.03$ μsec; total duration of exciting pulse 3 μsec, see Fig. 21), whence we estimated the velocity of motion of the domain; for a sample of length 1 mm the domain velocity was found to be $(3.5 \pm 0.4) \cdot 10^5$ cm/sec. This result is in good agreement with the velocity of sound in ZnS, equal to $3.4 \cdot 10^5$ cm/sec [45, 46]. We recall that the critical electron drift velocity corresponding to E_{cr} is $(3.6 \pm 0.3) \cdot 10^5$ cm/sec. Such ex- cellent agreement indicates that the observed electrical instability is acoustoelectric.

As a final demonstration that electric domains moving through the bulk of the crystal are responsible for electroluminescence in our case, we simultaneously displayed on the screen of a high-speed double-beam oscilloscope the waveform of the current through the crys- tal and the glow coming from different parts of the sample.

Figure 22a shows such an oscillogram for the case of a completely uncovered sample (light from all parts of the sample falls upon the photomultiplier input). It can be seen that the observed current and brightness oscillations are in phase. In Fig. 22b the near-cathode region of the crystal is covered, so that light from this region does not enter the photomulti- plier; it can be seen that the brightness oscillations are shifted toward the end of the current oscillations. When the region near the anode of the crystal is covered (Fig. 22c), the bright- ness oscillations are shifted towards the beginning of the current oscillations. It can be seen from these oscillograms that the amplitude of the electroluminescent brightness oscillations

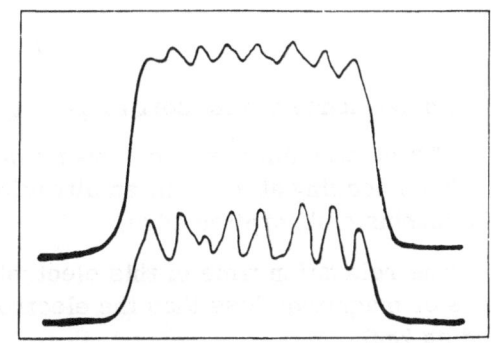

Fig. 21. Oscillograms of current through sam- ple (top) and electroluminescent brightness (bottom). Oscillograms taken simultaneously in a single sweep.

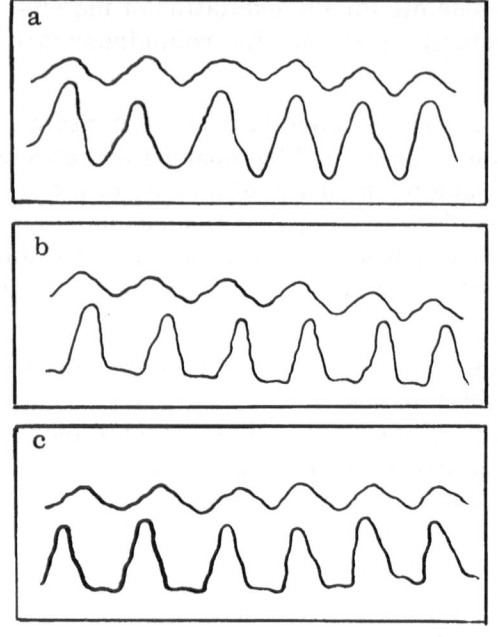

Fig. 22. Oscillograms of current through sample (top) and electroluminescent brightness from different parts of sample (bottom). Oscillograms taken simultaneously on a double-beam oscilloscope in a single sweep. (a) Entire crystal exposed; (b) near-cathode part of crystal covered; (c) near-anode part of crystal covered.

is quite unaffected by whether the entire sample is uncovered or whether the photomultiplier receives emission from only a part of the sample. The only change is in the duration of the brightness oscillations (partially covering the samples reduces the duration). Furthermore, partially covering a sample increased the dark region between oscillations, and also, as mentioned above, the brightness oscillations were displaced relative to the current oscillations.

All these results are consistent with the hypothesis that it is not the whole crystal that radiates but only a relatively narrow part of it, this luminescing region, like an electric domain, originating at the cathode and moving towards, the anode, where it disappears. The fact that the luminescing region moves along the sample at a velocity of ~10^5 cm sec in conjunction with the inertia of the eye means that in visual observation under the microscope the emission appears to be uniformly distributed over the volume and synchronous.

CONCLUSION

In summary, we have observed a low-voltage prebreakdown electroluminescence from ZnS single crystals devoid of internal and contact barriers. The electroluminescence is observed at voltages corresponding to mean fields in the crystal of ~10^3 V/cm.

The dependence of electroluminescent brightness on applied voltage has the form

$$B = B_0 \exp\left(-b/\sqrt{U}\right).$$

A B(U) dependence of this sort is quite typical for prebreakdown electroluminescence of ZnS.

The electroluminescence spectrum contains, besides a blue band peaking at 465 nm and a red band peaking at 700 nm, an ultraviolet band that extends down to 330 nm, which is close to the intrinsic absorption edge.

The relaxation time of this electroluminescence $\tau_{\text{relax}} \lesssim 0.2$ μsec, which is two to three orders of magnitude less than the electroluminescence relaxation time of powdered luminophors based on ZnS.

Our investigations disclosed a negative differential conductivity effect and the presence of electrical instability involving the creation, motion along the crystal, and decay of field-concentrating electric domains.

A connection is established between the observed electroluminescence and electrical instability. The emission is found to come not from the whole volume of the crystal but only from a relatively narrow part of it. The luminescing region of the crystal, just like a domain, nucleates near the cathode and then moves towards the anode, where it decays. Due to the high velocity of displacement of the domains, ~10^5 cm/sec, the visual impression on the eye is that the entire crystal luminesces uniformly and synchronously. The domain velocity and the critical electron drift velocity corresponding to the voltage at which electrical instability appears are found to be in good agreement with the velocity of sound in ZnS, indicating that the observed instability is acoustoelectric.

We suggest that the low-voltage electroluminescence is produced as a result of the concentration of electric field in the domain up to values sufficient to excite electroluminescence by electrical breakdown processes.

The relatively low working voltages, the spectral composition of the emission, and the frequency characteristics of the observed electroluminescence may be of practical value in the construction of low-voltage high-speed sources of visible and ultraviolet light.

In conclusion, it is our pleasant duty to thank A. V. Lavrov for kindly supplying the crystals, V. A. Chikhacheva for her assistance with the work, and I. K. Vereshchagin and V. A. Chuenkov for a discussion of the results and valuable remarks.

LITERATURE CITED

1. M. Cardona and G. Harbeke, Phys. Rev., A137:1467 (1965).
2. Yu. V. Bochkov and A. N. Georgobiani, Tr. FIAN, 50:60 (1970).
3. A. N. Georgobiani and V. I. Steblin, Tr. FIAN, 50:27 (1970).
4. A. N. Georgobiani, M. B. Kotljarevsky, V. N. Zlobin, P. A. Todua, Yu. P. Generalov, and B. P. Dementev, Mater. Res. Bull., 8:893 (1973).
5. G. S. Pekar', N. B. Luk'yanchikova, Hoang Mi Shin, and M. K. Sheinkman, Pis'ma Zh. Éksp. Teor. Fiz., 19:513 (1974).
6. A. N. Georgobiani, Usp. Fiz. Nauk, 113:129 (1974).
7. A. N. Georgobiani and P. A. Todua, Kratk. Soobshch. Fiz., No. 5, p. 26 (1970).
8. V. A. Chuenkov, Fiz. Tverd. Tela, 2:200 (1959).
9. V. A. Chuenkov, Izv. AN SSSR, Ser. Fiz., 20:1550 (1956).
10. V. Frantz, Dielectric Breakdown [Russian translation], IL, Moscow (1961).
11. L. V. Keldysh, Zh. Éksp. Teor. Fiz., 37:713 (1959).
12. H. Watanabe, Jpn. J. Appl. Phys., 5:12 (1966).
13. H. Lozikowski, Czech. J. Phys., B13:164 (1963).
14. V. A. Chapnin, Candidate's Dissertation, Physics Institute, Academy of Sciences of the USSR (1969).
15. H. Watanabe, T. Chikamura, and M. Wada, Jpn. J. Appl. Phys., 13:357 (1974).
16. N. G. Basov, O. V. Bogdankevich, and A. G. Devyatkov, Zh. Éksp. Teor. Fiz., 47:1588 (1964).
17. V. S. Vavilov, É. L. Nolle, and V. D. Egorov, Fiz. Tverd. Tela, 7:749 (1965).
18. É. L. Nolle, V. S. Vavilov, G. P. Golubev, and V. S. Mashtakov, Fiz. Tverd. Tela, 8:286 (1966).
19. C. E. Hurwitz, Appl. Phys. Lett., 9:116 (1966).
20. O. V. Bogdankevich, M. M. Zverev, A. I. Krasilnikov, and A. N. Pechenov, Phys. Status Solidi, 19:K5 (1967).

21. J. Bille, B. M. Kramer, P. Reimers, W. Ruppel, and R. Stille, Phys. Status Solidi, 36:K71 (1969).
22. O. V. Bogdankevich, N. A. Borisov, A. N. Georgobiani, V. B. Gutan, B. M. Lavrushin, O. V. Matveev, E. I. Panasyuk, V. F. Pevtsov, and N. L. Poletaev, Kvant. Élektron., 2:2231 (1975).
23. Yu. P. Chukova, Tr. FIAN, 37:149 (1966).
24. V. L. Bonch-Bruevich, I. P. Zvyagin, and A. G. Mironov, Domain Electrical Instability in Semiconductors [in Russian], Izd. Nauka, Moscow (1972).
25. B. K. Ridley, Proc. Phys. Soc., 82:954 (1963).
26. B. K. Ridley and T. B. Watkins, Proc. Phys. Soc., 78:293 (1961).
27. J. B. Gunn, IBM J. Res. Dev., 8:141 (1964).
28. Yu. N. Berozashvili, Candidate's Dissertation, Physics Institute, Academy of Sciences of the USSR (1967).
29. I. Yamashita, I. Ishiguro, and T. Tanaka, Jpn. J. Appl. Phys., 4:470 (1965).
30. N. G. Zhdanova, M. S. Kagan, and S. G. Kalashnikov, Fiz. Tverd. Tela, 8:788 (1966).
31. A. G. Foyt, R. E. Halsted, and W. Paul, Phys. Rev. Lett., 16:55 (1966).
32. A. R. Hutson, J. H. McFee, and D. L. White, Phys. Rev. Lett., 7:237 (1961).
33. M. E. Gertsenshtein and V. I. Pustovoit, Radiotekh. Élektron., 7:1009 (1962).
34. V. I. Pustovoit, Usp. Fiz. Nauk, 97:257 (1969).
35. N. G. Zhdanova, M. S. Kagan, and S. G. Kalashnikov, Fiz. Tverd. Tela, 8:744 (1966).
36. V. L. Bonch-Bruevich, Fiz. Tverd. Tela, 6:2041 (1964).
37. V. I. Pustovoit, S. G. Kalashnikov, and G. S. Pado, Fiz. Tekh. Poluprovodn., 3:832 (1969).
38. Sinclair S. Yee, Appl. Phys. Lett., 9:10 (1966).
39. L. A. Sorokina, Candidate's Dissertation, Physics Institute, Academy of Sciences of the USSR (1966).
40. R. W. Smith, Phys. Rev. Lett., 9:87 (1962).
41. Physical Acoustics. Vol. 4A: Applications to Quantum and Solid State Physics, W. P. Mason, ed., Academic Press, New York (1966).
42. A. R. Moore, Phys. Rev. Lett., 12:47 (1964).
43. C. Fischler-Hazoni and F. Williams, Phys. Rev., 139:A583 (1965).
44. M. Aven and C. A. Mead, Appl. Phys. Lett., 7:8 (1965).
45. D. Berlincourt, H. Jaffe, and L. R. Shiozava, Phys. Rev., 129:1009 (1963).
46. W. E. Spear and P. G. Le Comber, Phys. Rev. Lett., 13:434 (1964).
47. A. Sadhu and K. C. Kao, Solid State Commun., 8:2013 (1970).
48. W. J. Fleming and J. E. Rowe, J. Appl. Phys., 42:2041 (1971).
49. A. N. Georgobiani, A. I. Blazhevich, Yu. V. Ozerov, E. I. Panasyuk, P. A. Todua, and H. Friedrich, Izv. Akad. Nauk SSSR, Ser. Fiz., 37:415 (1973).
50. A. N. Georgobiani, A. I. Blazhevich, H. Friedrich, Yu. V. Ozerov, E. I. Panasyuk, and P. A. Todua, in: Luminescence of Crystals, Molecules, and Solutions, F. Williams, ed., Plenum Press, New York (1973), p. 239.
51. F. Matossi, K. Lentwein, and G. Schmidt, Z. Naturforsch., 21a:461 (1966).
52. Physics and Chemistry of II-IV Compounds [Russian translation], Izd. Mir, Moscow (1970).
53. A. N. Georgobiani and P. A. Todua, J. Lumin., 5:14 (1972).
54. A. N. Georgobiani and P. A. Todua, Kratk. Soobshch. Fiz., No. 6, p. 28 (1971).
55. A. N. Georgobiani, Tr. FIAN, 23:3 (1963).
56. I. K. Vereshchagin, Electroluminescence of Crystals [in Russian], Izd. Nauka, Moscow (1974).
57. V. E. Oranovskii, Izv. Akad. Nauk SSSR, Ser. Fiz., 25:516 (1961).
58. S. S. Yee, Solid State Electron., 10:1015 (1967).

LUMINESCENCE EXCITATION SPECTRUM FOR EXCITATION IN THE REGION OF FUNDAMENTAL ABSORPTION, AND THE ELECTRONIC STRUCTURE OF ALUMINUM NITRIDE

V. V. Mikhailin, V. E. Oranovskii, S. I. Pachesova, and M. V. Fok

A detailed study is made of the structure of the impurity-luminescence excitation spectrum of AlN for excitation in the range 6-12 eV. In addition to broad bands, the excitation spectrum is found to contain a series of narrow lines of width less than 0.05 eV. These lines can be ascribed to excitons belonging to different band extrema. These excitons carry away the excitation energy deep into the crystal, thereby diminishing the excitation density, which leads in the given crystals to an increase in the luminescence yield. The phonon replicas of these lines corresponding to phonon energies of 114, 92, and 57 meV are also found. The positions of the narrow lines and the broad bands in the excitation spectrum are found to be in good agreement with the theoretical calculations of Jones and Lettington. It is also shown that the broad-band peaking at 8.7 eV is due to the appearance in the AlN lattice of corundum nucleations.

In the present work we investigate the band structure of aluminum nitride crystals and the nature of the electronic transitions excited by photons in the region of the fundamental absorption. The band structure of crystals is usually studied in terms of the spectral variation of the reflection coefficient R. This method has many shortcomings, however. In the first place it involves laborious calculations, starting from the Kramers-Kronig relations, of the absorption coefficient K, as it is the latter that is directly associated with the band structure. The absorption coefficient K determined in this manner is very imprecise, as the directly measured quantity, R, can vary by a factor of not more than 5-6 (from 15-20% in the region of transparency to almost 100% in the region of large K) whereas the corresponding variation of the quantity of interest, K, can be more than 5-6 orders of magnitude (for example, from 0.1 to 10^5 cm^{-1}).

Another shortcoming of this method is that the reflection spectrum, like the absorption spectrum, tells us nothing about where the energy of the photons that were absorbed by the crystal has gone. Further, peaks in the absorption (or reflection) spectra may be connected not only with features of the band structure, but also with absorption due to defects or impurities in the crystal lattice if they are present in sufficient numbers or to the appearance of collective excitations, which are likewise ignored in band structure calculations.

Our investigations were accordingly based on the luminescence excitation method, which (especially when combined with reflectivity data) yields much more information. In particular by studying the luminescence excitation spectrum in a specific spectral region, we can assess what incident photon energy is transmitted most efficiently to which luminescent center. If the integrated-luminescence excitation spectrum is compared with the reflection spectrum, information can be obtained on the spectral dependence of the luminescence yield and, consequently, not only on radiative transitions but on radiationless transitions as well.

Let us consider this matter in more detail. The luminescence excitation spectrum coincides with the product of the ordinates of the absorption spectrum and the luminescence yield spectrum of the body. If the absorption coefficient of the exciting light K is small, so that absorption has still no effect on the reflection coefficient R, and if the crystal thickness is sufficiently small that the product

$$Kd \ll 1, \tag{1}$$

then the absorbance is proportional to Kd. The spectral dependence of the reflection coefficient R may be ignored, since in the present case R is determined by the refractive index, which, far from the absorption band, is little dependent on wavelength. If the luminescence quantum yield is also independent of the wavelength of the exciting light, then the luminescence excitation spectrum repeats the absorption spectrum.

For the same value of K but for a thick sample for which

$$Kd \gg 1, \tag{2}$$

we see that the excitation spectrum replicates the spectral dependence of the luminescence yield as the absorbance is then almost constant, being equal simply to $1 - R$, which, as already noted, is little dependent on wavelength.

In the case in which we are interested, however, the absorption coefficient K is sufficiently large that it has an effect on R, so that R ceases to be a constant. The large value of K means that inequality (2) is fulfilled, indeed with "reserve." Accordingly, for a constant luminescence quantum yield, the excitation spectrum repeats the spectral dependence of $1 - R$. In other words, the excitation spectrum again proves to be connected with the absorption spectrum, as was the case for small K and inequality (1) satisfied. However, since R increases with increasing K, maxima of absorption now coincide not with maxima but with minima in the excitation spectrum.

Dependence of the luminescence quantum yield on the wavelength of the exciting light causes the excitation spectrum to deviate from the spectral dependence of $1 - R$. These deviations may be so great that, for example, a maximum may appear on the excitation spectrum instead of a minimum. In other cases, however, the minima may remain on their places but may become much deeper than one would expect from the spectral dependence of $1 - R$.

Let us consider how these variations can be explained. We assume that photon multiplication does not occur, i.e., we assume that each quantum of exciting light leads to the creation of not more than one quantum of luminescence. This means that we are restricting the discussion to exciting photons of energy less than twice the bandgap. The factors below, which determine the dependence of the recombination-luminescence quantum yield on the wavelength of the exciting light, are relevant under these conditions.

1. Dependence of Luminescence Quantum Yield on Excitation Density. A dependence of this kind often arises due to competition between recombination centers of two or more sorts when at least one of them involves recombination without radiation. In this case the spectral dependence of the quantum yield depicts the absorption spectrum, since the same number of exciting quanta are absorbed for different K in layers of different

thickness and so create an excitation of different density. Here, in turn, two cases are possible: (a) If the yield increases with excitation density (i.e., if the dependence of the brightness on excitation intensity is superlinear), then maxima of absorption will correspond to maxima of quantum yield. (b) If, on the other hand, the yield falls with increasing excitation density (i.e., the brightness varies sublinearly with excitation intensity), then the spectral dependence of the quantum yield represents the absorption spectrum as in a mirror, so to speak, as absorption maxima correspond to yield minima.

2. Presence of Radiationless Recombination Centers on the Surface of the Crystal. These centers can originate from imperfection of the crystal lattice in the surface layers, and also as a result of absorption of foreign atoms onto the surface of the crystal. Under such conditions the luminescence quantum yield also proves to be dependent on the absorption coefficient, since the thinner the layer in which the exciting light is absorbed, the closer to the surface will electrons and holes be created, and the probability that they will come out onto the surface and recombine there without emission is correspondingly greater. Accordingly, in this case maxima in the absorption spectrum correspond to minima of luminescence quantum yield.

3. Appearance of Excitons, Plasmons, and Other Mobile Forms of Collective Excitation. On its own this can have no effect on the luminescence yield; it can, however, have a considerable effect on the luminescence yield in the presence of the factors 1 and 2 above. The fact is that these collective excitations, thanks to their mobility, are capable of carrying energy deep into the body of the crystal, thereby reducing the excitation density. Accordingly, they influence the quantum yield in exactly the same way as a reduction of the absorption coefficient. The only difference is that the appearance of excitons and plasmons manifests itself as an increase in the absorption. Accordingly, maxima in the absorption spectrum must, under the present conditions, correspond to yield extrema that are opposite what was observed in cases 1 and 2 above. In other words (a) if the yield increases with increasing excitation density, the excitation of impurity recombination luminescence must be a minimum in the vicinity of exciton and plasmon absorption peaks; (b) if the yield decreases with increasing excitation density, then these peaks must correspond to maxima of excitation; (c) if radiationless recombination centers are present on the surface of the crystal, then in this case too exciton absorption peaks must correspond to maxima of excitation of recombination luminescence.

4. Large Concentration of Impurity Centers or Lattice Defects Having Strong Absorption. If they themselves are luminescence centers, or if they transfer energy to luminescence centers with greater efficiency than for the absorption of exciting light by the crystal lattice itself, then maxima of the absorption band of these impurity centers (defects) will also be maxima of luminescence quantum yield. This factor shows up particularly strongly if the absorbing centers are located near the surface of the crystal.

In order to go over from the spectral dependence of the luminescence quantum yield to the excitation spectrum, allowance must also be made for the spectral dependence of the absorbance, which, as was mentioned above, is described in all four cases by the quantity $1 - R$. This quantity has minima near maxima of K. The corresponding minima of quantum yield are thus somewhat enhanced in the excitation spectrum, and maxima somewhat eroded.

The spectral dependence of the luminescence quantum yield is therefore obtained by dividing the ordinates of the excitation spectrum by the corresponding values of $1 - R$. However, if a maximum in the excitation spectrum does not correspond to a minimum of R, it can be asserted without this operation that a maximum of luminescence quantum yield occurs in this region. Several factors can be responsible for the appearance of such yield maxima, and in each specific case we have to establish which of them plays the dominant role.

Method of Measurement

The reflection spectra were measured as described in [1]. The luminescence excitation spectra were measured for excitation by light from a windowless hydrogen or helium lamp for exciting photon energies in the range 5.5-14 eV, and also in the range 6-36 eV using synchrotron UV radiation. In both cases the exciting radiation was passed through a concave diffraction grating (with a radius of curvature of 1 m) having 1200 lines/mm and operating in the first order. For operation in the range up to 6.7 eV, the second diffraction order was suppressed by placing a quartz plate in front of the monochromator outlet slit.

The emission was recorded using FÉU-51 or FÉU-51 photomultipliers with electronically modulated gain. The signal from the photomultiplier was fed to a narrow-band amplifier and was recorded by a digital voltmeter. Exciting light scattered by the sample was prevented from entering the photomultiplier by placing in front of the latter a BS-12 light filter transparent in the range 3.5-1.2 eV (360-1000 nm); in some cases a further light filter was placed in front of the photomultiplier to remove the green region peaking near 2.4 eV (530 nm).

As a rule, the excitation spectra were measured "point by point." The duration of excitation at each point varied from 3 to 10 min depending on the wavelength of the exciting light, and was sufficient to enable the luminescent brightness to reach its steady value. The luminescence efficiency $\eta(h\nu)$ at each wavelength was determined from the results of four separate measurements:

1. A sodium salicyilate screen, having a constant (to within 10%) luminescence quantum yield in the range of exciting photon energies from 30 to 3.6 eV [2] was placed in front of the monochromator outlet slit. The intensity of the exciting flux together with the background of scattered light, J_{eb}, was measured.
2. The monochromator grating was moved aside and the intensity of the background of scattered light, J_b, was measured.
3. The sample was mounted in front of the monochromator inlet slit in place of the sodium salicylate screen. The luminescence excited by the scattered light, J_{lb}, was measured.
4. The grating was introduced. The luminescence J_{leb} induced by the exciting light and the scattered light simultaneously was measured.

The luminescence excitation efficiency (in relative units) was computed using the equation

$$\eta = (J_{leb} - J_{lb})/(J_{eb} - J_b). \tag{3}$$

This equation is strictly valid only if the luminescent intensity varies linearly with excitation intensity. Superlinear and sublinear variations give values of η that are slightly too large and too small respectively. However, so long as $J_{leb} \gg J_b$, the error in the determination of η is not large.

A nonlinear variation of luminescent intensity with excitation intensity means that η is itself dependent on excitation intensity. Accordingly, to find the spectral dependence of η in this case, the excitation intensity (expressed in the number of incident quanta) must be kept constant. To this end, the width of the monochromator slit must be varied in such a manner as to keep J_{eb} constant.

The investigated AlN samples showed a nonlinear (more accurately, a sublinear) dependence of luminescent intensity on excitation intensity (Fig. 1). Accordingly, in our experiments, the number of photons incident upon the sample had to be kept constant, which was done by varying not only the width of the monochromator slit but the discharge current through the lamp as well. The minimum spectral width of the slit then amounted to 0.02-0.05 eV for photon

I$_{lum}$, rel. units

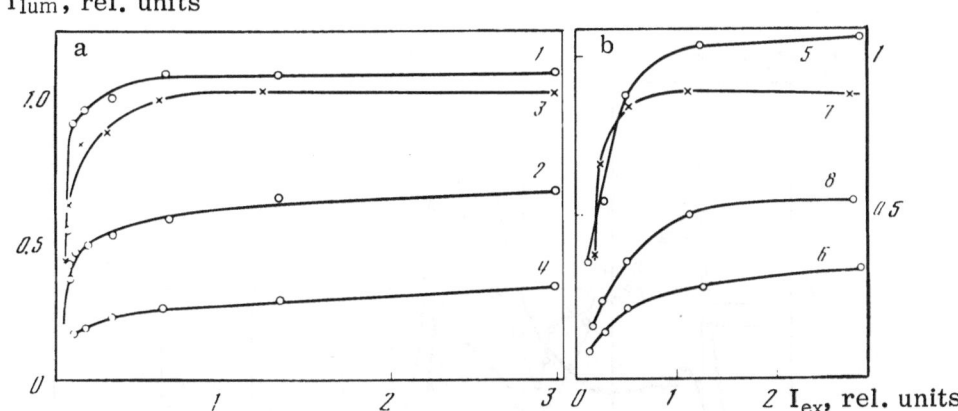

Fig. 1. Dependence of luminescent intensity of AlN on excitation intensity (relative units). (a) Oxygen concentration 1.84 mole%. Curves 1 and 2 correspond to excitation $h\nu$ = 6.2 eV; luminescence: curve 1, λ = 500 nm; curve 2, λ = 380 nm. Curves 3 and 4 correspond to excitation $h\nu$ = 11.47 eV; luminescence: curve 3, λ = 500 nm; curve 4, λ = 380 nm. (b) Oxygen concentration 11.1 mole %. Curves 5 and 6 correspond to excitation $h\nu$ = 6.2 eV; luminescence curve 5, λ = 500 nm; curve 6, λ = 380 nm. Curves 7 and 8 correspond to excitation $h\nu$ = 8.7 eV; luminescence: curve 7, λ = 500 nm; curve 8, λ = 380 nm.

energies in the range 5.5–12 eV and to 0.07 eV for the range 12–14 eV. In the investigation of broad bands in the reflection spectrum, the slit width was greater by a factor of 4–10; during continuous recording of the spectrum, when the phonon structure was being investigated, the slit width was around 1.5 times less.

In addition to the "point-by-point" measurements, those parts of the excitation spectrum where the spectrum of the hydrogen or helium lamp did not contain any narrow lines were continuously recorded at a narrow monochromator slit width. This enabled us, without having to introduce any corrections, to observe narrow excitation bands, which could easily be missed in the point-by-point measurements.

Samples made from AlN powder were prepared for the measurements in the form of compressed layers of thickness around 0.1 mm. Single-crystal samples were used only for excitation with synchrotron UV radiation. The single crystals, which had the shape of prismatic "pencils," were stacked parallel to each other so as to form an almost continuous area. They were oriented so that their C axis was parallel to the monochromator slit and perpendicular to the plane of the electric vector of the synchrotron radiation (which was approximately 85% polarized).

The investigated samples were not specially doped with anything; however, they all contained not less than 0.5% (by weight) of oxygen, which was incorporated into the AlN lattice during the process of synthesis.

Broad Bands in the Excitation Spectrum

First of all, we investigated the broad bands that could be noticed even for the comparatively sparsely distributed experimental points obtained with a wide slit. Crystalline powders of aluminum nitride (molar concentration of oxygen from 1.84 to 11.1%) (Fig. 2) and single

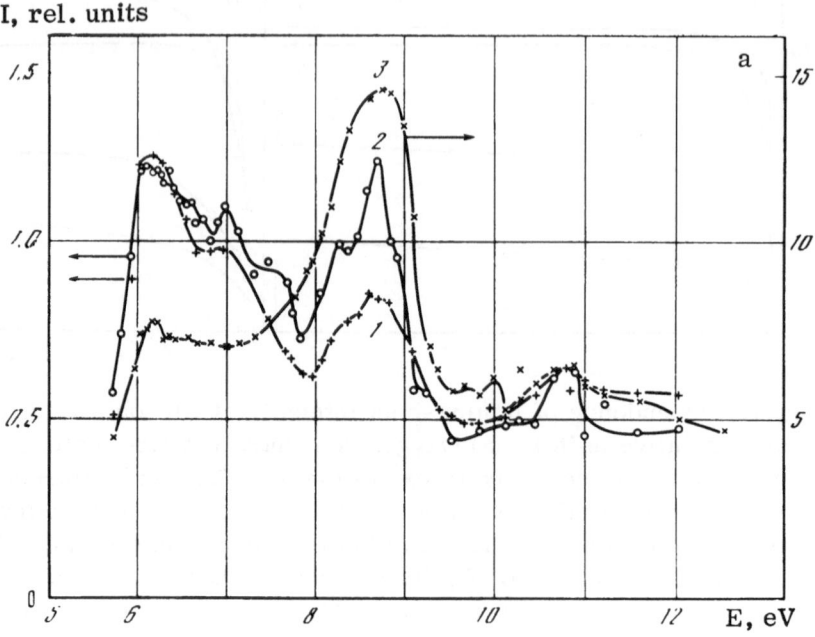

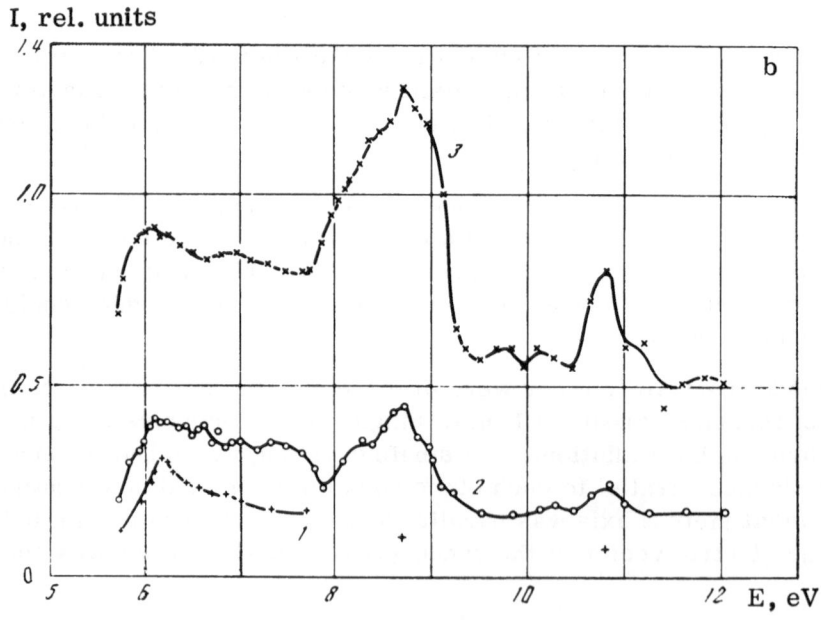

Fig. 2. Luminescence excitation spectra of crystalline powders of AlN with different molar concentrations of oxygen. (a) Integrated luminescence (2.3 eV ≤ hν_{em} ≤ 3.5 eV); (b) the green component (hν_{em} = 2.3 eV). Oxygen concentration: (1) 1.84%; (2) 3.2%; (3) 11.1%; T = room temperature; spectral width of slit 0.08–0.2 eV; arrows indicate ordinate scale.

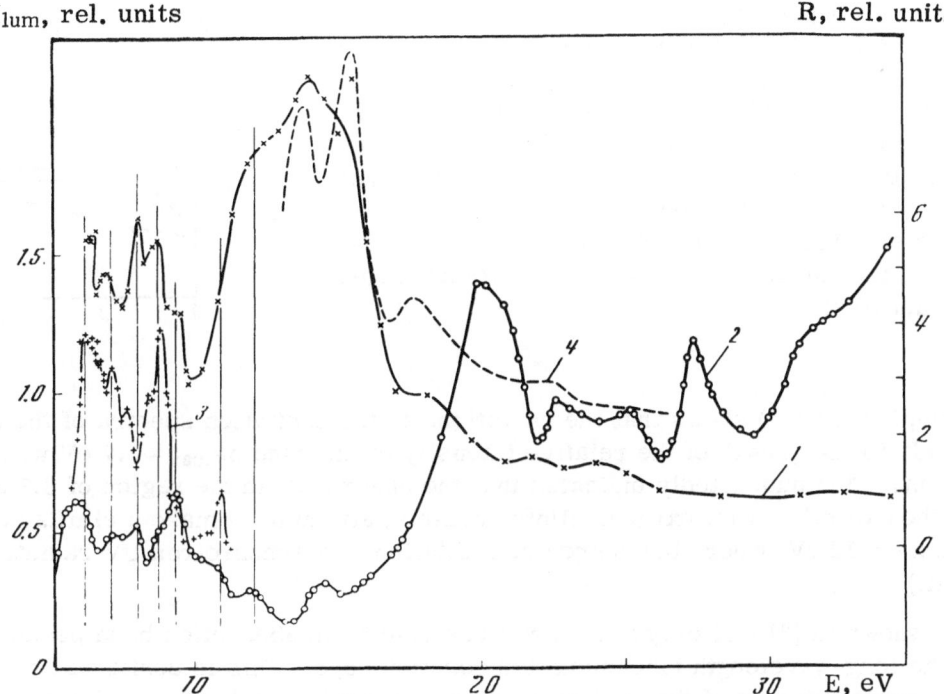

Fig. 3. Comparison of excitation and reflection spectra of AlN. Curves
1. and 2 are reflection and excitation spectra of one and the same batch
of AlN single crystals; spectral width of slit is 0.3 eV for $h\nu \leq 10$ eV and
1 eV for shorter wavelengths. Curve 3 is excitation spectrum of crystalline powder (curve 2 from Fig. 2a). Curve 4 is reflection spectrum of a
large single crystal (spectral width of slit 0.3 eV). All the curves are
plotted in different relative units, as the scale along the vertical axis depends on the setup and geometry of the sample. The vertical lines are
drawn to show that reflection maxima correspond, in most cases, not to
minima of the excitation spectrum but to maxima.

crystals (Fig. 3) were studied. It can be seen from these figures that the excitation spectrum
of aluminum nitride consists of a whole series of bands. In particular, in the range up to 12 eV,
the spectrum contains three well-resolved bands ($h\nu_{max}$ = 6.2, 8.7, 10.8 eV) and one less clearly expressed band ($h\nu_{max}$ = 7.0 eV). The short-wavelength part of the excitation spectrum also
contains a number of bands, for example, with peaks at 20 and 27 eV. The excitation peaks at
7.0, 7.9, 8.7, and 9.2 eV, and probably also the peak at 6.2 eV, correspond to peaks or
"shoulders" (but never to minima) in the reflection spectrum. On the other hand, the clear
excitation peaks in the range 17-27 eV correspond neither to peaks nor to minima in the reflection spectrum. This indicates that the proportion of the incident radiation entering the
crystal does not change by much and is masked by the variation of some other factor, determining the spectral dependence of the luminescence yield.

It is immediately obvious from a study of the figures that the excitation spectra of different samples differ considerably from each other. Thus, the 7.0 eV band is most apparent
in the spectrum of single crystals; the 8.7 eV band, on the other hand, has in the single-crystal spectrum only the form of a shoulder on the edge of another band ($h\nu_{max}$ = 9.2 eV), which,
in turn (like the band $h\nu_{max}$ = 7.9 eV), is hardly visible at all in the excitation spectrum of
powders. A more detailed comparison of the amplitudes of individual bands in the spectra of

Fig. 4. Dependence on oxygen concentration of normalized (with respect to band $h\nu_{max} = 6.2$ eV) intensities of integrated-luminescence excitation bands. (1) $h\nu_{max} = 7.0$ eV; (2) $h\nu_{max} = 8.7$ eV; (3) $h\nu_{max} = 10.5$ eV. The horizontal straight line corresponds to the band $h\nu_{max} = 6.2$ eV (as the positions of the peaks of the bands are themselves slightly dependent on oxygen concentration, the values cited here are certain averaged quantities).

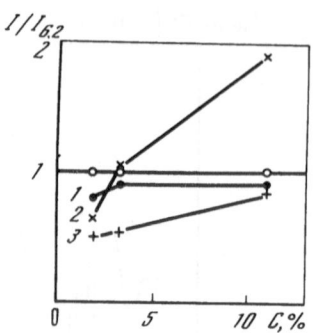

different samples (Fig. 4) shows that the variations of the excitation spectra of the samples are due mainly to the growth of the relative intensity of the band $h\nu_{max} = 8.7$ eV with increasing oxygen content. This undoubtedly indicates that the absorption in the region of 8.7 eV is connected somehow or other with oxygen. (Unfortunately, attempts to make a similar comparison in the range $h\nu > 12$ eV, where the source of excitation was synchrotron UV radiation, were unsuccessful.)

It was shown in [3] that oxygen in AlN gives rise to an absorption band peaking at 4.5–4.8 eV next to the long-wavelength fundamental absorption edge. This is confirmed by the study of the long-wavelength part of the excitation spectrum made in [4]. It was shown in this paper that AlN luminescence in the region 2.3–3.6 eV is excited in the range 3.2–4.9 eV, both spectra consisting of a superposition of several bands belonging to different luminescence centers containing one or two oxygen atoms.

From a comparison of Figs. 2a and 2b we see that the excitation band at 8.7 eV shows up at large oxygen concentrations even more strongly in the integrated luminescence than in the green. It follows that in this region blue-emitting luminescence centers are efficiently excited as well as green-emitting centers. It is hardly likely that different luminescence centers would have a second absorption band in exactly the same place. Furthermore, each act of excitation in this band must involve the dissipation at each center of not less than 5 eV, since the energy of the luminescence photons does not exceed 3.7 eV. In other words, almost two-thirds of the energy of the exciting photon must be dissipated. Stokes losses of this magnitude within a luminescence center are not usually observed. It is more likely that these losses occur outside the luminescence center, in the transfer to it of excitation energy from some more complex oxygen formation located in the AlN lattice.

The oxygen concentration in the investigated samples is considerable. Clusters of oxygen atoms containing three to four atoms each must accordingly exist in the samples at neighboring lattice sites.

As is well known, impurities are usually displaced out of the regular lattice into disturbed regions, as a result of which their concentration near dislocations may be much greater than the mean concentration taken over the volume of the sample. The concentration of luminescence centers will be correspondingly increased. Accordingly, in such regions of the crystal, excitation energy will be transferred to luminescence centers more efficiently.

This transfer probably occurs in several stages. First of all, energy is transferred from an oxygen formation to the AlN lattice with the formation of a "hot" electron—hole pair. This transition is allowed both in energy and in momentum, since the absorption peak (8.7 eV) ascribed to oxygen formations is located beyond the AlN fundamental absorption edge, and this means that the AlN band structure contains many pairs of levels that are in resonance with an electronic transition in the oxygen formation. Forbiddenness in momentum is removed, since the transition involves not only the crystal lattice but a local formation (oxygen cluster) as well, which is capable of taking up the difference in the momenta.

The electron—hole pair produced in this manner is rapidly thermalized and almost immediately one of the charge carriers is trapped by a luminescence center. Some of the excitation energy is thereupon converted into heat. The large concentration of luminescence centers near an oxygen inclusion means that they can compete here more successfully with radiationless recombination centers than far from the inclusion. Thus the yield is maximum in the vicinity of the absorption of the oxygen inclusion. Subsequent recombination at a luminescence center is also accompanied by a release of heat. (This recombination occurs much more slowly than the first stages of energy transfer, as the charge carriers are hindered on capture centers.)

In this manner, in accordance with our hypothesis, the excess 5 eV is converted into heat in several stages in almost the same way as in the absorption of exciting light in the host material, when the absorbed photon directly induces an interband transition. The difference consists in "spatial" (and not "energy") circumstances, and it is this that leads to the increase of the luminescence yield.

It now remains for us to explain why the absorption peak for the oxygen clusters should occur just at 8.7 eV. This energy corresponds to the bandgap of α-Al_2O_3 (corundum) [5-10]. As is well known, one of the peaks in the luminescence excitation spectrum always lies on the fundamental absorption edge. It is thus natural to suppose that this peak in the excitation spectrum is associated with absorption in corundum grains. However, as experiment shows, if corundum separates out in the form of a separate phase, the luminescent brightness of the sample is markedly reduced. Furthermore, the corundum phase separates out at still greater oxygen concentrations. In the investigated samples, however, the corundum phase was not observed. This forces us to ascribe the 8.7 eV band to corundum nucleations that have not yet manifested themselves in the form of a separate phase.

Let us consider how corundum nucleations can occur in the crystal lattice of aluminum nitride. Both corundum and aluminum nitride have a hexagonal crystal lattice in which the anions are almost close-packed (the distortion in corundum is slightly greater than in AlN). The distance between nearest unlike atoms in corundum [11] is 1.97 and 1.86 A, and in aluminum nitride it is 1.92 and 1.88 Å [12]. The distance between nearest oxygen atoms in corundum equals 2.49 and 2.87 Å, whereas in α-Al_2O_3 (which has a cubic lattice) it is equal to 4.18 Å. The distance between nearest nitrogen atoms in aluminum nitride is equal to 3.11 Å, i.e., it lies within those limits between which the distance between oxygen atoms in aluminum oxide crystals can vary. The main difference between the structures of corundum and aluminum nitride lies in the disposition of the aluminum atoms: in AlN they occupy tetrahedral vacancies; in corundum, octahedral. Further, in AlN aluminum atoms occupy one-half of the tetrahedral vacancies; in corundum, octahedral. Further, in AlN aluminum atoms occupy one-half of the tetrahedral vacancies, and in corundum two-thirds of the octahedral (Fig. 5).

In this manner, all that is required for a corundum nucleation to appear in the crystal lattice of AlN containing oxygen impurity is that a few nitrogen atoms be substituted by oxygen alongside an aluminum vacancy and that one of the nearest aluminum atoms be displaced from a tetrahedral to an octahedral vacancy.

A straightforward calculation shows that for a molar concentration of oxygen of from 1.84 to 11.1% (the range obtaining in our samples), the molar concentration of octahedra in which three or more nitrogen atoms are substituted by oxygen varies respectively from $1 \cdot 10^{-2}$ to 2%. If, however, near a dislocation, the local oxygen concentration increases by a factor of

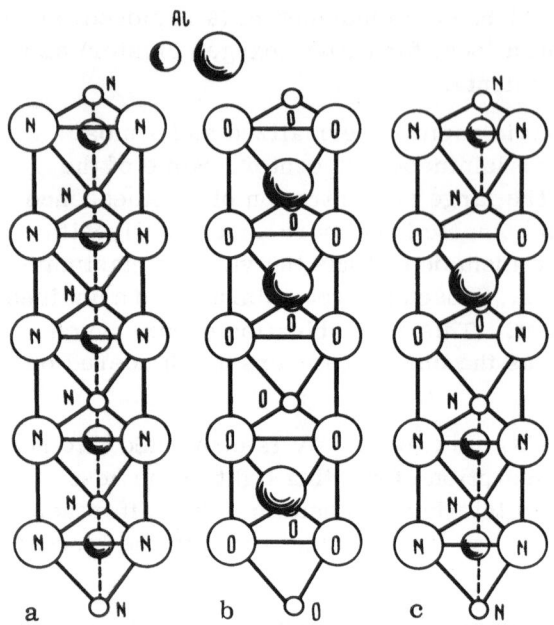

Fig. 5. Comparison of crystal lattice of aluminum nitride and corundum. Part of lattice of (a) aluminum nitride, (b) corundum, and (c) aluminum nitride with corundum nucleation.

only 2-3, then the number of such octahedra increases sharply and can reach 0.1-0.3% even at the minimum considered oxygen concentration. At this sort of concentration of corundum nucleations, their absorption can already manifest itself in the excitation spectrum in the form of a separate band, even on the background of the fundamental absorption. A factor promoting this is that oxygen centers are most likely to be located near the surface of the crystallites, since there are always more dislocations and defects there.

They are hardly likely, however, to form a continuous oxide film, since a repeated measurement of the oxygen concentration made 5 years after the samples were prepared showed that the oxygen concentration in them had increased in all by 0.1-0.4% (the fine crystalline powders were oxidized more strongly than the coarse). Even if we assume that a film 2-3 times as thick grew on the surface of the crystallites as they were being cooled after the preparation of the AlN powders, it would be able to substitute only a small proportion of the oxygen occurring in these crystallites. The thickness of such a film amounts in all to 30-40 Å. Also, at such thicknesses, films are rarely continuous. But even if a continuous film were formed, it would absorb only a few percent of the exciting light even at an absorption coefficient of 10^5 cm^{-1}. All this testifies that we are dealing not with surface but with bulk properties of aluminum nitride, although indeed the volume which we are investigating extends into the body of the crystal 1000 Å in all.

The remaining peaks in the investigated part of the excitation spectrum probably belong to the host material, i.e., to the aluminum nitride. A more detailed analysis is not possible without involving the fine structure of these peaks, and it is this that is the subject matter of the next section.

Fine Structure of the Excitation Spectrum

A more detailed study of the excitation spectra shows that their structure consists of more than just broad bands. In particular, they were found to contain a whole series of very narrow and intense excitation bands and also a less sharply expressed structure reminiscent of phonon structure.

I_lum, rel. units

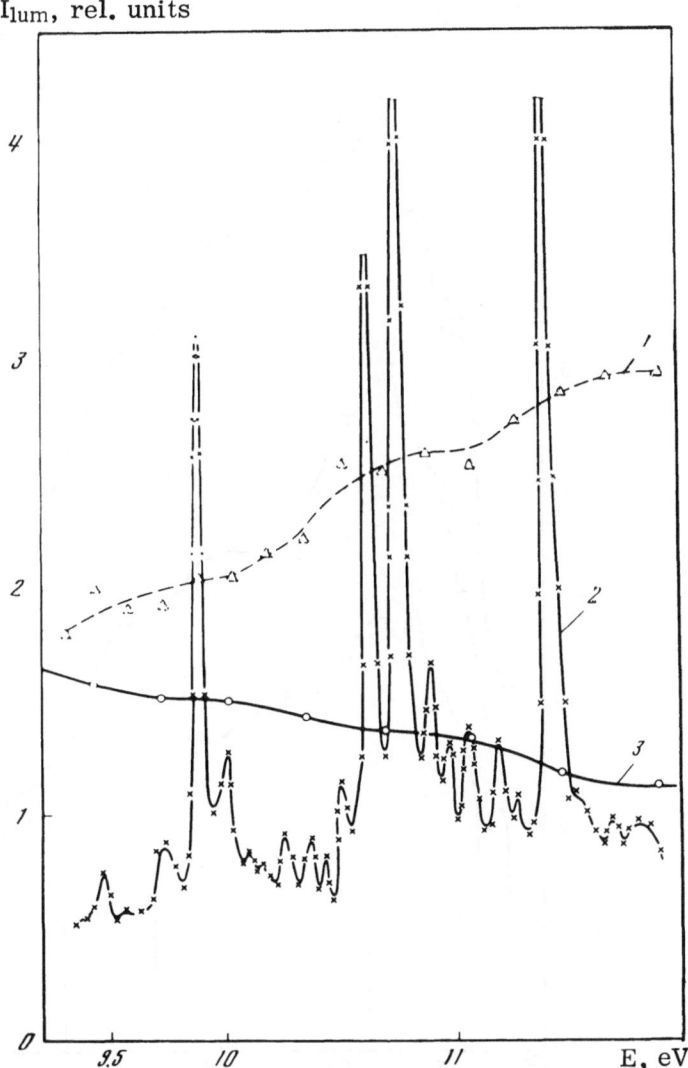

Fig. 6. A portion of the excitation spectrum of AlN.
(1) Curve measured "point by point" (spectral width
of monochromator slit 0.04 eV); (2) curve measured
in continuous recording (spectral width of slit 0.03
eV; the short-wavelength peaks have been cut off as
they extend beyond the limits of the figure; the heights
of the peaks corresponding to $h\nu$ = 9.86, 10.63, 10.79,
and 11.46 eV are respectively 3.2, 7.8, 11.3, and 6.3
rel. units); (3) spectrum of exciting source, measured
in continuous recording (it has no special features in
the range of interest). Curves 1-3 are expressed in
different arbitrary units.

   Figure 6 shows the most intense group of narrow bands, measured in continuous recording
with a narrow monochromator slit. They are characterized by a small halfwidth (less than
0.05 eV) and by a considerable asymmetry: the high-energy wing is almost twice as broad as
the low-energy wing (this asymmetry is not very noticeable in the figure as the horizontal scale
is not great enough; this scale is extended just the same, as the figure shows only a small part
of the measured excitation spectrum).

I_lum, rel. units

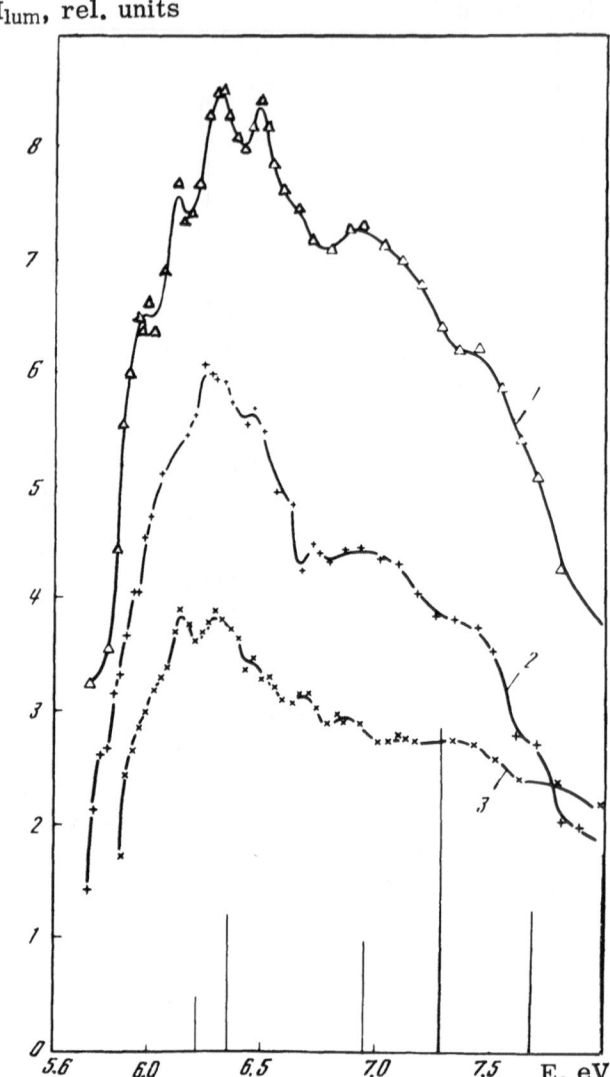

Fig. 7. Excitation spectra of three samples of aluminum nitride with different oxygen concentrations. (1) 6.04%; (2) 4.42%; (3) 3.21%. Spectral width of slit was varied from 0.02 eV (for $h\nu = 8$ eV). Source of exciting light is a hydrogen lamp. The vertical lines denote the positions and intensities of narrow peaks.

The small halfwidth of these bands means that they can easily be missed in "point-by-point" measurements, which is illustrated, for example, by curve 1 of this figure. Figure 7 shows, for three samples of different oxygen concentrations, a portion of the excitation spectrum measured "point-by-point." The vertical lines indicate narrow excitation bands. In the indicated scale along the horizontal axis, the thickness of these lines is only 2-3 times less than the width of the excitation bands which they represent. The lengths of the lines represent the intensities of the corresponding excitation bands.

An excitation spectrum with a complex structure like this cannot be explained solely by interband transitions, since the theory of interband transitions in the one-electron approximation predicts only broad bands devoid of fine structure. Also, the halfwidth of the narrow excitation bands is much less than what one would expect on the basis of the rate of variation of the density of states near a band extremum. Finally, the theory of interband transitions in the one-electron approximation in no way explains why the energy of light quanta corresponding to an absorption peak is more likely to be transferred to luminescence centers than the energy of quanta corresponding to an absorption minimum. In both the first case and the second the light entering the crystal is absorbed in it entirely. Accordingly, we suggest that the observed narrow excitation bands are due not to the appearance of electron−hole pairs, but to the appearance of some other elemental excitation.

The small value of the halfwidth of these narrow excitation bands suggests that neither can they be ascribed to separate absorbing centers in the AlN lattice, since the excitation bands on the fundamental absorption edge belonging to oxygen centers [4] and also the above-considered oxygen bands peaking at 8.7 eV have a halfwidth that is at least an order of magnitude greater than these narrow lines. Accordingly, we suggest that they belong to excitons of Wannier−Mott type. The electrons and holes bound into an exciton are usually assumed to correspond to the outermost band extrema (to the lowest minimum of the conduction band and the highest maximum of the valency band). Exciton absorption and emission are thus sought near the long-wavelength edge of the absorption due to interband transitions. However, excitons can also be produced when the electrons and holes belong to other band extrema, provided only that they have momenta that are equal in absolute magnitude (and opposite in direction).† They will then have an energy that is considerably greater than that of ordinary "edge" excitons.

The production of excitons upon absorption of the exciting light may, as was mentioned above, give rise to a more efficient energy transfer to luminescence centers (see p. 43, case 3b and c). Since the luminescent intensity of our samples depends sublinearly on excitation intensity, it follows that we have to do with case 3b and not 3c. That surface centers of radiationless recombination play a small role is further confirmed by the fact that the absorption of exciting light by oxygen centers, which we assume to be located near the surface, leads to the appearance of an excitation peak (at $h\nu = 8.7$ eV) and not a minimum.

If our ideas on the nature of the narrow bands in the excitation spectrum are correct, then each broad "interband" absorption band corresponding to transitions between particular band extrema should possess its own narrow exciton absorption peak, associated with the same band extrema and shifted slightly toward longer wavelengths relative to the interband absorption peak. And this exciton peak should correspond to a coincident peak in the excitation spectrum. Exciton peaks far beyond the fundamental absorption edge have been observed in the absorption spectra of alkali-halide crystals [13] and also in the absorption and reflection spectra of Cu_2O [14].

The band structure of aluminum nitride was first computed by Hejda [15] by the orthogonalized plane-wave (OPW) method. He found that the width of the forbidden band is connected with direct transitions at the point $k = 0$ and that at the top of the valence band the point Γ_1 lies above Γ_6 by approximately 0.1 eV. However, these calculations gave a value of $E_G = 2.35$ eV for the width of the forbidden band; this contrasts with the experimental value of 5.74 eV for

---

† As is well known, this requirement follows from momentum conservation: as the momentum of a photon is small, it follows that the momentum of the exciton produced upon its absorption must also be small.

the electric vector parallel to the C axis, and 5.88 eV for the electric vector perpendicular to the C axis [3]. In a subsequent paper by Hejda and co-workers [16] it was shown that even in a more refined form the OPW method does not give more satisfactory agreement with experiment (E_G = 3.91 eV).

The most exact calculation of the band structure of aluminum nitride was made by Jones and Lettington [17]. In this calculation the authors did not utilize a single free parameter which could be adjusted to give better agreement with experiment. The only experimental information utilized by the authors was crystal structure data: a = 3.111 Å; c/a = 1.6. The free parameter of the theory (which allowed for the energy shift of the closed electron shells of nitrogen) was found by the authors from independent experiments; it was selected so that the calculated width of the forbidden band of gallium nitride coincided with the experimental value. No subsequent adjustment of this parameter was made in the bandgap calculation for aluminum nitride. These calculations confirmed Hejda's results on the order in which the bands are located at k = 0 and gave a value of the bandgap (of around 6 eV) that was much more consistent with experiment. Here, however, it must be kept in mind that in the initial calculation the distance between the points Γ_{6V} and Γ_{1V} of the valence band proved to be much too large. In the opinion of the authors of [17] this happened because the point Γ_1 is calculated insufficiently reliably (as was noted in [17], this point is the most sensitive to slight variations of the free parameter). The authors therefore calculated the splitting $\Gamma_{6V} - \Gamma_{1V}$ independently by two methods and obtained, instead of 0.75 eV, only 0.11 and 0.27 eV. If this circumstance is taken into account, and the point Γ_{1V} is shifted so that the splitting of the valence band works out to be 0.19 eV (the mean of 0.11 and 0.27 eV), then the width of the forbidden band is found to be 5.87 and 6.06 eV (for the two polarizations), which is very close to the experimental results of [3] cited above.

According to our results, however, the excitation spectrum peaks at slightly higher energies (6.2-6.4 eV). This gives still better agreement with theory, since, when comparing the results of a band structure calculation with experiment, it must be remembered that the width of the forbidden band corresponds only to the start of absorption. The absorption always peaks in fact at higher photon energies. Accordingly, the $\Gamma_{6V} - \Gamma_{1V}$ transition with a calculated energy of E_T = 6.06 eV is completely compatible with the experimental excitation peak at hν = 6.3 eV.

The fact that such good coincidence is achieved without adjusting the free parameter and the fact that the structure of the AlN crystal lattice is taken into account in the calculations give grounds for hoping that the data on the remaining band extrema of AlN obtained in these calculations are just as reliable.

Going on ahead for the moment, we note that the broad excitation spectrum we are talking about has a fine structure. This structure we regard as phonon structure with principal lines at 5.85 and 6.16 eV. Bearing in mind that these lines were observed at 110°K, while kinks at hν = 5.74 and 5.88 eV in the absorption spectrum were observed at room temperature, it can be said that the positions of the lines and kinks are in good relative agreement since cooling through nearly 200° is quite likely to result in a 0.2-0.3 eV increase in the width of the forbidden band. However, lines that are so narrow as to permit observation of their phonon structure are hardly likely to belong to interband transitions. They are more likely to belong to excitons, in which case the corresponding kinks in the absorption spectrum must also be regarded as associated with excitons. Now, as is well known, the energy of excitons varies with temperature in almost the same manner as the width of the forbidden band. The correspondence established above between the lines in the excitation spectrum and the kinks in the absorption spectrum should thus also hold in the event that they are of excitonic origin.

Assuming that this is so, the energies corresponding to these kinks can be compared with the theoretical value of the width of the forbidden band (5.87 and 6.06 eV for room temperature). Hence we find that the exciton binding energy equals 0.15-0.20 eV.

A theoretical estimate of the binding energy of a hydrogenlike exciton situated in the lowest state gives, as is well known,

$$E_{ex} = (\mu^* e^4)/(2\hbar^2 \varepsilon^2), \tag{4}$$

where μ^* is the reduced excitonic mass, ε is the dielectric constant of the crystal, and e is the electronic charge. From [18-20], the dielectric constant equals 4.67, 4.68, or 4.84. Taking the mean of these values ($\varepsilon = 4.73$) and setting $\mu^* = 0.5m_e$ (where m_e is the free-electron mass), we find that $E_{ex} = 0.3$ eV. Within the limits of accuracy of the calculation, this agrees with the "semiempirical" results cited above (to make the coincidence exact it is sufficient, for example, to take $\mu^* = 0.25m_e$).

From the excitation and reflection spectra given above it can also be seen that the reflection spectrum has a peak at $h\nu = 8.6$ eV and a shoulder at 9-9.3 eV, while in this range of energies the excitation spectrum has an exciton peak at $h\nu = 8.53$ eV. We attribute it to $U_{3V}-U_{3C}$ transitions ($E_T = 8.5$ eV), and the shoulder to $U_{4V}-U_{3C}$ transitions ($E_T = 8.9$ eV). This determination of the transitions is consistent with the results of an analysis of the optical spectra of ZnS, CdS, and CdSe, compounds also of wurtzite structure. It was shown in [21] that the structure of the spectra of all these compounds is due to transitions of the same type. As in AlN, in the compounds ZnS, CdS, and CdSe the onset of direct transitions (peak E_0) corresponds to transitions $\Gamma_{6,1V}-\Gamma_{1C}$; the next peak E_1 (in order of increasing energy) corresponds to $U_{3,4V}-U_{3C}$ and $\Gamma_{5V}-\Gamma_{3C}$ transitions; peak E_2 corresponds to transitions at the M, K, and H corner points of the Brillouin zone.[†] The type and the location of the critical points associated with these transitions also proved to be the same in ZnS, CdS, and CdSe. Accordingly, we would expect that in AlN, as in ZnS, CdS, and CdSe, the region of $U_{3,4V}-U_{3C}$ transitions lies around the U axis of the Brillouin zone with critical points of type M on the U axis and in its vicinity within the zone in the direction towards Γ [21]. In AlN, as in the other wurtzite-lattice compounds investigated in [22], the energy of these transitions is around 2.5 eV above E_G.

The excitation spectrum of single crystals obtained using synchrotron UV radiation (which is, as is well known, almost completely polarized; in our case perpendicularly to the C axis) contains a distinct peak at $h\nu = 9.2$ eV. This peak can hardly be seen at all, however, in the excitation spectrum of powders. It follows that the degree of polarization of this transition (as in ZnS, CdS, CdSe) is very considerable. We attribute the peak at $h\nu = 9.2$ eV to the transition $\Gamma_{5V}-\Gamma_{3C}$ ($E_T = 9.3$ eV), which must also be polarized perpendicularly to the C axis. The direction of polarization follows from the selection rules for transitions in the region of the Γ-Δ-A axis. The critical points of type M_1 associated with the $\Gamma_{5V}-\Gamma_{3C}$ transition are located at Γ and on the Δ axis [21].

Unfortunately, the energy values for the remaining transitions are not cited in [17]. They can be obtained only from the graph, which results, of course, in a certain error. Energy values found in this manner will be indicated below by an asterisk. The reflection peak with $h\nu = 7.9$ eV and the excitation band with $h\nu = 7.8$ eV and also the excitonic peak with $h\nu = 7.69$ eV can be assigned to the transition $\Gamma_{1V}-\Gamma_{3C}$ ($E_T = 7.5^*$ eV) or $\Gamma_{6V}-\Gamma_{3C}$ ($E_T = 7.7^*$ eV). In an ideal lattice of wurtzite type these transitions are forbidden by the selection rules. However, as the parameters of the AlN lattice differ significantly from the parameters of the ideal wurtzite lattice (the ratio $c/a = 1.600$ instead of 1.633), the selection rules may be expected to be relaxed.

[†] E_0, E_1, E_2, etc. is the notation proposed by Cardona and Harbeke [22] for the elements of the structure of the reflection spectra.

The next significant absorption peak for compounds of zinc blende structure is denoted E_2. It is connected with transitions in the region of points X $(X_{5V}-X_{1C})$ and K of the Brillouin zone. State X_1 in zinc blende is split into X_1 and X_3; state X_5 is also split on account of spin—orbit interaction. Two positions are possible in the Brillouin zone of the wurtzite lattice for points analogous to points X and K of zinc blende: near the point M in a region extending along the Σ axis, and near the point K in a region extending parallel to the Z direction. According to [21], the main contribution to this absorption band in crystals with wurtzite structure comes from $M_{2V}-M_{1C}$ transitions; lesser contributions come from the transitions $K_{2V}-K_{2C}$, $K_{3V}-K_{2C}$ and $H_{3V}-H_{3C}$.

The band $h\nu = 10.7$ eV, the most intense excitonic peak $h\nu = 10.79$ eV, and the peak $h\nu = 10.63$ eV in the excitation spectra of the samples we ascribe to transitions $M_{2V}-M_{1C}$ ($E_T = 11.0$ eV) or $K_{2V}-K_{2C}$ ($E_T = 10.9$ eV) and $K_{3V}-K_{2C}$ ($E_T = 10.7$ eV) respectively. The peak with $h\nu = 10.63$ eV is spaced from the peak $h\nu = 10.79$ eV by ~0.1 eV, which may be connected with a splitting of the valence band $K_{2V}-K_{3V}$. The peak observed at $h\nu = 9.86$ eV can probably be attributed to the transition $M_{4V}-M_{3C}$ ($E_T = 9.8$ eV); this means, however, that we would have to admit that this energy is calculated less precisely than the rest, as no place remains for the exciton binding energy in this case. The excitonic peak with $h\nu = 11.41$ eV is superposed on the broad band at 11.6 eV in the excitation spectrum and agrees with the position of the peak with $h\nu = 11.7$ eV observed in the reflection spectrum. This peak we attribute to the transition $U_{1V}-U_{1C}$ ($E_T = 11.6^*$ eV) in an X-like region near the point M. The regions of transitions which contribute to the band E_2 are the neighborhoods of the M, K, and H corner points of the

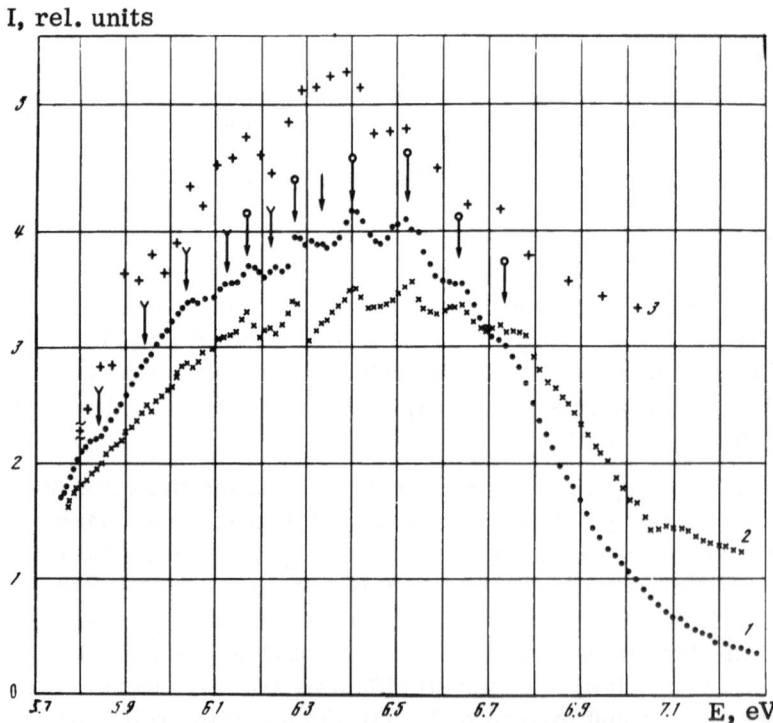

Fig. 8. Structure of excitation spectrum of AlN (11.1 mole % oxygen) for excitation on fundamental absorption edge. (1) T = 108°K; (2) T = 113°K; (3) T = 295°K. Spectral width of slit 12-15 meV. The arrows indicate the calculated positions of the peaks in the phonon series.

TABLE 1. Vibrational Structure of Band with $h\nu = 6.3$ at $110°$K

$h\nu_{meas} - h\nu_{calc}$, meV	$h\nu_{calc}$, eV ($h\nu_{ph} = 114$ meV)	$h\nu_{meas}$, eV	$h\nu_{calc}$, eV ($h\nu_{ph} = 92$ meV)	$h\nu_{meas} - h\nu_{calc}$, meV
		5.846	5.847	−1
		5.940	5.939	+1
		6.035	6.031	+4
		6.120	6.123	−3
−4	6.164	6.160		
		6.215	6.215	0
−10	6.278	6.268		
		6.32		
+3	6.392	6.395		
+9	6.506	6.515		
+5	6.620	6.625		
−1	6.734	6.733		

Brillouin zone with critical points of type M_2 located at the most highly symmetric points or on the boundary of the broad regions around them [21].

Transitions to other excited states begin in the more short-wavelength part of the spectrum, in particular, to the second conduction band. For example, the "shoulder" region at $h\nu = 12$-12.7 eV in the reflection spectrum and the flat peak at $h\nu = 12.5$ eV in the excitation spectrum probably correspond to the transition $\Gamma_{1,6V} - \Gamma'_{1,6C}$ ($E_T = 12.75*$ eV).

More detailed measurements of the reflection spectra disclosed three further peaks, at 13.7, 5.3, and 17.5 eV (see Fig. 3, curve 2). However, as the number of possible transitions increases rapidly with photon energy, on which, to boot, structure connected with excitons is superposed, it follows that an interpretation of the spectra in the short-wavelength region requires both more detailed measurements and also more exact theoretical calculations. Accordingly, in the present paper we restrict ourselves to the range up to 12 eV, and ew now go on to consider the phonon structure of the spectra.

Phonon Structure of the Excitation Spectra

Structure similar to phonon structure was observed in almost all parts of the excitation spectrum of AlN that were measured in detail. Unfortunately, however, we did not manage to measure the entire excitation spectrum with sufficient detail. Two series of peaks with different repetition periods were discovered in the region of 6 eV (Fig. 8). The energies of the corresponding phonons were determined so as to make the calculated system of peaks differ as little as possible from the measured system. We consider that the error in the determination of this energy does not exceed 2 meV. (This error value is arrived at as follows: agreement between the calculated and the experimental positions of the peaks is noticeably worsened by varying the phonon energy by ±1 meV; the error is augmented by a second millielectron volt to allow for measurement uncertainty, which amounted to 3-4 meV for an individual peak but which was correspondingly less for the entire system.) The results obtained† are cited in Table 1.

In this table the quantity $h\nu_{meas}$ is obtained as the mean of two measurements: at 108 and 113°K. These figures are accurate to within a few units of the last digit. As can be seen from

---

† It should be kept in mind that these data on the phonon structure are obtained at a temperature of around 110°K, whereas the rest of the measurements were made mainly at room temperature. This led to a certain discrepancy in the positions of the peaks obtained in different experiments.

the table, the mean-square deviation of the calculated position of a peak from the measured position does not exceed 3 meV for the more long-wavelength system of peaks and 7 meV for the more short-wavelength system, which amounts to less than 10% of the distance between peaks. This testifies to the quite strict constancy of the distance between the observed peaks, a feature characteristic of phonon replicas. In addition to the peaks cited in Table 1, all the measured spectra contained a barely discernable peak at $h\nu$ = 6.32 eV; its phonon replicas could not be differentiated, however, probably because of their small intensity.

The energy spacing between the peaks, equal to 114 meV, is in good agreement with the energy of zone-center LO phonons, which have [23, 24] an energy of 113.6 or 112.8 meV and belong to the Γ_5 and Γ_1 representations [25].

The phonon energy of 92 meV is close to the LO-phonon energy of 91.4 and 91.6 meV found in [23, 26]. In these papers the phonons were ascribed to the B-zone boundary. In our case, however, we are probably dealing with zone-center phonons, as they interact efficiently with excitons belonging to the Γ_1 and Γ_6 representations; interaction of zone-center optical transitions with zone-boundary phonons is improbable.

Let us elucidate to which of the nine optical-phonon branches occurring in the lattice vibration spectrum of AlN phonons with energy 92 meV belong. The absence of calculated phonon dispersion curves and neutron scattering data for AlN means that the interpretation of the observed frequencies will be of a hypothetical character. However, with the aid of the known symmetry of the corresponding transitions in AlN and utilizing the calculated phonon dispersion curves for compounds with a similar form of binding as AlN and the same crystal

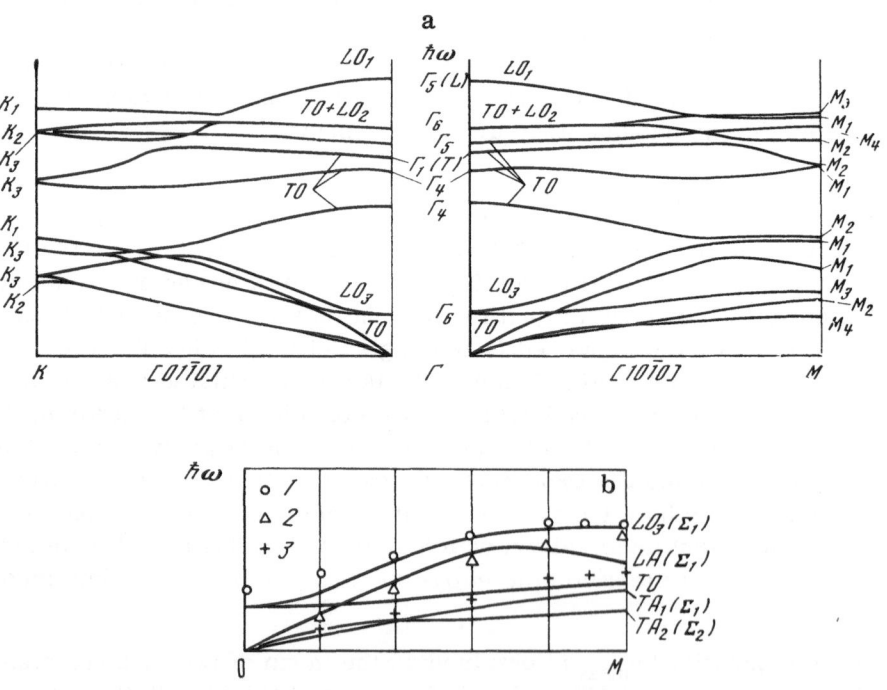

Fig. 9. Phonon dispersion curves for crystals with wurtzite lattice. (a) Theoretical curves for CdS; (b) comparison of calculated [25] phonon dispersion curves for CdS (the solid curves) with neutron scattering data for BeO (1) LO (Σ_1) mode; (2) LA (Σ_1) mode; (3) TA (Σ_1) mode.

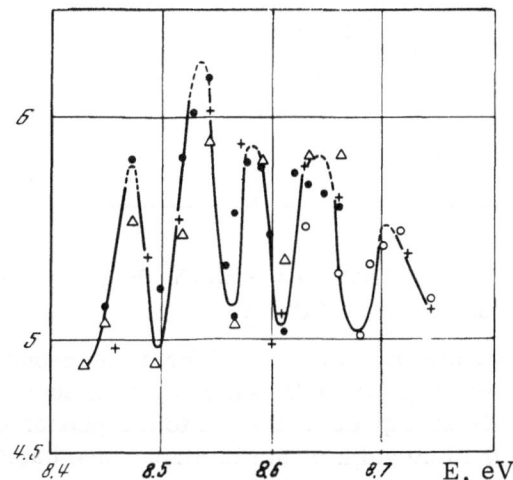

Fig. 10. Structure of excitation spectrum of AlN near 8.5 eV. Temperature; optical width of slit 20 meV. The different symbols refer to measurements made at different times and on different samples prepared from power with 11.1 mole% oxygen concentration.

structure, we can, with reasonable reliability, determine the symmetry and the mode of the vibration participating in the electronic transition in AlN.

The phonon dispersion curves for the wurtzite lattice were calculated in [25] allowing for short-range and long-range forces modeling the covalent and ionic parts of the binding in the crystal (Fig. 9a). Although the calculation was carried out for CdS, the disposition and symmetry of the phonon branches for other crystals with a wurtzite lattice and mixed binding (ionic and covalent) may well be qualitatively the same. We checked this by comparing the experimentally measured phonon dispersion curves for the wurtzite-structure compound BeO [27] with the calculated curves for CdS [25]. As can be seen from Fig. 9b, calculation and experiment are in qualitative agreement.

According to experimental data for the wurtzite-structure compounds, CdS, ZnS [20], and ZnO [28], it is LO phonons that interact most strongly with optical transitions. The phonon dispersion curves were calculated in [25] for CdS, but, as mentioned above, we utilized these results for AlN as well. It follows from these calculations that at the zone center the phonons LO_1, LO_2, and LO_3 (listed in order of decreasing energy) belong, in the direction A-Γ to the representations Γ_1, Γ_4, and Γ_4 respectively, and to the representations Γ_5, Γ_6, and Γ_6 in the directions M-Γ and K-Γ.

As was mentioned above, the more energetic phonon (114 meV) belongs to the representations Γ_5 and Γ_1. This corresponds to the LO_1 phonon branch. It can also be seen from Fig. 9 that the TO + LO_2 branch is situated at the points K and Γ almost at exactly the same level, although in the interval between these points it is split into branches LO_2 and TO that are spaced from each other by quite a noticeable amount. Accordingly, the phonons with energies 91.4 and 91.6 meV found in [22, 24] must be assigned to the points K or M on the zone boundary, despite the fact that the phonons with almost the same energy (92 meV) found by us belong to the zone center and have Γ_6 symmetry (branch LO_2).[†]

[†] The equality of the energy of the branches $LO_2(K_2) \approx TO(K_2) \approx TO(K_3)$ and also of the branches $LO_3(K_3) \approx TA_1(K_2) \approx TA_2(K_3)$ explains why it is sometimes possible (e.g., in [22]) to interpret the absorption spectrum in the multiphonon region with the aid of a total of only six zone-boundary phonon frequencies, whereas the lattice vibration spectrum of AlN contains 12 branches.

TABLE 2. Phonon Structure in Region 8.4–8.7 eV

$h\nu_{meas}$, eV	$h\nu_{calc}$, eV ($h\nu_{ph}$ = 57 meV)	$h\nu_{meas}-h\nu_{calc}$ meV	$h\nu_{meas}$, eV	$h\nu_{calc}$, eV ($h\nu_{ph}$ = 57 meV)	$h\nu_{meas}-h\nu_{calc}$ meV
8.470	8.473	−3	8.640	8.644	−4
8.535	8.530	+5	8.708	8.701	+7
8.580	8.587	−7			

We consider now the phonon replicas of the narrow band with $h\nu$ = 8.53 eV (Fig. 10). The measured and calculated positions of these peaks are compared in Table 2.

It can be seen from Table 2 that the phonon energy amounts to 57 meV for a mean-square deviation of the calculated peak position from the measured of 5.4 meV, which again amounts to less than 10% of the distance between the peaks. This energy corresponds to the phonon energy determined from the temperature dependence of the halfwidth of the luminescence band with λ_{max} = 1 μm and from an analysis of the shape of this band [29]. The mean weighted value of the phonon energy obtained by these two methods is 56.1 meV. The obtained energy value also agrees with the phonon energies of 55.3 and 55.1 meV determined from an analysis of the absorption spectra in the infrared [23, 25].

The 8.53 eV band, as mentioned above, is due to $U_{3V}-U_{3C}$ transitions from the region of the Brillouin zone lying on the M–U–L axis within the zone. $U_{3V}-U_{3C}$ transitions are direct,[†] the energy spacing between peaks corresponding to the energy of a single phonon. Hence it follows that the phonons that interact with an electronic transition belong to the zone center or to the region around the center [30, 31]. Comparing the calculated phonon energies corresponding to the calculated dispersion curves with the phonon energy observed in experiment, we can ascribe this phonon to a TO(LO) mode[‡] of Γ_4 symmetry (see Fig. 9). A "pure" (i.e., on the U axis) electronic transition $U_{3V}-U_{3C}$ involving phonons of Γ_4 symmetry is forbidden by the selection rules. Nonetheless, besides the critical point on the U axis (of type M_0), a critical point (of type M_1) is located near the axis within the zone [21], and as a result electronic transitions associated with this region and close to $U_{3V}-U_{3C}$ in energy can occur with participation of Γ_4 phonons from the vicinity of Γ, especially in the Σ direction. The 57-meV Γ_4 phonon frequency has hitherto not been observed, as this mode is active neither in IR absorption nor in Raman scattering.

We also managed to analyze the phonon replicas of the narrow band with energy 10.79 eV (Fig. 11). The phonon energy in this case was found to be 97 ($\pm$ 2) meV. This energy differs from the energy of all phonons hitherto considered by an amount greater than the error in the determination of the energy of the corresponding modes; it therefore requires another interpretation.

The 10.79 eV band is due to transitions in the region of the M–K axis. It can be shown using the selection rules [21] that phonons of Γ_6 symmetry are allowed in the corresponding

[†] Indirect Γ–H transitions have an energy that is almost coincident with the energy of $U_{3V}-U_{3C}$ transitions. However, as is well known, the transition dipole moment between points of different symmetry type is zero or negligibly small [21]. Accordingly, if there are no additional considerations favoring indirect transitions (for example, direct transitions forbidden in the dipole–dipole approximation), the optical structure with corresponding energy is ascribed to direct transitions.

[‡] In the Γ–A direction it is an LO mode and in the M–K direction a TO mode.

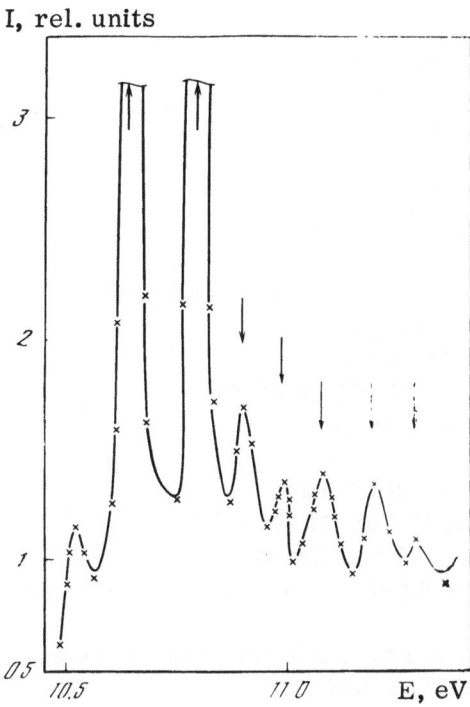

Fig. 11. Structure of excitation spectrum of AlN in region 10.5–11.5 eV. Temperature: room temperature; optical width of slit 35 meV; oxygen concentration 11.1%; the arrows indicate the calculated positions of the peaks.

optical transitions. Since the region of the transitions is a broad region around M, optical transitions close to $M_{2V} - M_{1C}$ in energy will occur with participation of phonons from the vicinity of Γ (particularly in the Σ direction). It can be seen from the dispersion curves that the frequency 97 meV definitely does not relate to the $LO_3(\Gamma_6)$ branch, as $LO_3(\Gamma_6)$ is much lower than $LO_2(\Gamma_6) = 92$ meV.

In this manner, the frequency $h\nu = 97$ meV can be regarded as a mode of the LO_2 branch with an energy slightly greater than that of zone-center modes.

A second exciton band with an energy of 10.63 eV can also be seen in Fig. 11. Its phonon replicas from the short-wavelength side are probably masked by the phonon replicas of the 10.79 eV band. There is only a single peak on the short-wavelength side; it can be interpreted as the phonon replica of the 10.63 eV band corresponding to the absorption of a phonon. It is confirmed by a maximum at 110 meV. However, as there is only one such peak, the energy of the corresponding phonons cannot be determined with sufficient accuracy. Accordingly we make no attempt to determine their symmetry.

Conclusion

The discovery of exciton bands and their phonon replicas in the region of the fundamental absorption opens up new possibilities for studying the band structure of aluminum nitride. In all probability, similar bands should also be observed in the luminescence excitation spectra of other semiconductors. A necessary condition for this, however, is that the luminescent brightness of the semiconductor should depend in a sharply sublinear manner on excitation intensity, or that surface recombination should lower the luminescence yield significantly.

LITERATURE CITED

1.   B. N. Meleshkin, V. V. Mikhailin, V. E. Oranovskii, G. A. Orekhanov, J. Pastrnák, S. Pachesová, A. S. Salamatov, M. V. Fok, and A. S. Yarov, Tr. FIAN, 80:140 (1975).

2.   F. S. Johnson, K. Watanabe, and R. Tousey, J. Opt. Soc. Am. 41:702 (1951).
3.   J. Pastrnák and Roskovcová, Phys. Status Solidi, 26:591 (1968).
4.   J. Pastrnák, S. Pacesová, and L. Roskovcová, Czech. J. Phys., B24:1149 (1974).
5.   E. T. Arakawa and M. W. Williams, J. Phys. Chem. Solids, 29:735 (1968).
6.   A. N. Laufeur, S. A. Pirog, and S. R. Mc. Nesby, J. Opt. Soc. Am., 55:64 (1965).
7.   L. Dunkelman, W. B. Fowler, and N. Hennes, Appl. Opt., 1:695 (1962).
8.   T. A. Chubb and H. Friedman, Rev. Sci. Instrum., 26:493 (1955).
9.   D. F. Heath and P. A. Sacher, Appl. Opt., 5:937 (1966).
10.   J. G. Lemonier, M. Priel, and S. Robin, C. R. Acad. Sci. Paris, 257:1508 (1963).
11.   Ruby and Saphire [in Russian], Izd. Nauka, Moscow (1974).
12.   G. A. Jeffrey and G. S. Parry, J. Chem. Phys., 23:406 (1955).
13.   J. E. Eby, K. J. Teegarden, D. B. Dutton, Phys. Rev., 116:1099 (1959).
14.   S. Brams and S. Nikitine, Solid State Commun., 3:209 (1965).
15.   B. Hejda, Phys. Status Solidi, 32:407 (1969).
16.   B. Hejda and K. Hauptmanova, Phys. Status Solidi, 36:K95 (1969).
17.   D. Jones and A. H. Lettington, Solid State Commun., 11:701 (1972).
18.   J. A. Kasaki and M. Mashimoto, Solid State Commun., 5:851 (1967).
19.   G. A. Cox, D. O. Cummis, K. Kawabe, and K. H. Tredhold, J. Phys. Chem. Solids, 28:543 (1967).
20.   D. C. Reynolds, C. W. Litton, T. C. Collins, and Y. S. Park, International Conference on Phonons, Flammarion Sciences, Paris, Rennes (1971), p. 367.
21.   T. K. Bergstresser and M. L. Cohen, Phys. Rev., 164:1069 (1967).
22.   M. Cardona and G. Harbeke, Phys. Rev., 137A:1467 (1975).
23.   A. T. Collins, E. C. Linghtowlers, and P. J. Dean, Phys. Rev., 158:833 (1967).
24.   O. Brafman, G. Lengyel, S. S. Mitra, P. J. Plendl, and L. S. Mansur, Solid State Commun., 6:523 (1968).
25.   M. Nusimovici and J. L. Birman, Phys. Rev., 156:925 (1967).
26.   J. Pastrnák and B. Hejda, Phys. Status Solidi, 35:941,953 (1969).
27.   G. L. Ostheller, R. E. Schmunk, R. M. Burgger, and R. J. Kearney, in: Neutron Inelastic Scattering, Vol. 1, Unipub, New York (1968), p. 315.
28.   S. S. Mitra and J. I. Bryant, J. Phys., 26:610 (1965).
29.   J. Rosa, Czech. J. Phys., B22:851 (1972).
30.   M. Nusimovici, J. Phys. (Paris), 26:689 (1965).
31.   M. Nusimovici, Ann. Phys., 4:97 (1969).

INVESTIGATION OF CONDENSATION OF EXCITONS IN GERMANIUM BY LIGHT-SCATTERING METHOD

N. N. Sibel'din

The basic properties of the condensed phase of excitons are studied and the possibilities of the light-scattering method of investigating the phase transition in which an exciton gas condenses into electron—hole drops are assessed. A very sensitive setup developed especially to study light scattering electron—hole drops is described. The kinetics of exciton condensation is studied via the measured dependence of the dimensions and concentration of the electron—hole drops on temperature, intensity, and method of excitation. The density and surface tension of the electron—hole liquid are measured. A model of exciton condensation is considered in which liquid-phase nucleations are formed as a result of fluctuations of the exciton density and in which the nucleations grow due to the diffusion of excitons toward their surface.

L. V. Keldysh [1] suggested in 1968 that a system of Wannier—Mott excitons in a semiconductor may undergo a phase transition similar to the first-order phase transition in a gas—liquid system. When the exciton density in the semiconductor exceeds a certain threshold value corresponding to the given temperature (for temperatures below the critical temperature), the excitons condense into "liquid" droplets, the electrons and holes in which are collectivized and bound by the forces of internal interaction.

Research on this new physical phenomenon was first commenced in the Soviet Union [2-7] but is now intensively conducted in many laboratories throughout the world. The number of papers devoted to the problem of exciton condensation is steadily increasing. Interest in this problem, which is probably one of the central problems in solid state physics [8], arises from a number of factors. First, electron—hole drops (EHD) possess a whole string of unusual properties: a high density for a small average (taken over the volume of the crystal) density of the nonequilibrium carriers, a high mobility in nonuniform fields, ans so on. An electron—hole liquid may also exhibit superconductivity or superfluidity [3]. Second, in the study of electron—hole liquids we encounter a situation that is, to a certain extent, exceptional for the many-body problem [9]. Since all the characteristics of electrons and holes in semiconducting crystals are quite well known (as opposed to the case of metals), it is possible to make a careful comparison of experimental data with theory and, in this manner, check the various approximations of the theory of metals.

The significance of the problem of exciton condensation goes, however, well beyond the limits of solid physics. Investigation of the system consisting of an exciton gas and an electron—hole liquid using the latest experimental methods yields much valuable information on the behavior of matter near the critical point, a topic of great interest at the present time (see [10], for example). The small value of the surface tension of an electron—hole liquid [11-15] means that, under certain conditions, exciton condensation may occur as a result of hetero-phase fluctuations [16] of the density of the exciton gas, a situation that is probably very rare in nature. We mention a further problem that can be modeled with the aid of a system of high-density excitons. Calculations show [17] that in germanium, in magnetic fields of the order of 10^5 Oe, the metallic phase consists of two-dimensional electron−hole cylinders. An analogous situation obtains at the surface of neutron stars, although the fields there are of the order of 10^{12} Oe.

Such investigations are also of practical importance in view of the increasing number of devices of optoelectronics that operate at high levels of optical excitation. For example, esti-mates show that it may be possible to observe stimulated emission from electron−hole drops [18].

Considerable progress has been made in recent years in understanding the processes underlying the condensation of excitons and the nature of the metallic state. An enormous amount of theoretical and experimental work has been done on this problem. Experimental studies have been carried out over a wide range of conditions using the most diverse methods, based, for example, on the following: photoluminescence [5, 19-38]; absorption and emission in the far infrared [6, 39, 40]; photoconductivity in dc [27, 41, 42] and at microwave frequen-cies [43-51]; and scattering of infrared radiation [52-58]. Electron−hole drops have been studied under conditions of uniaxial compression [7, 23, 26, 28, 56, 59-64]; in strong magnetic fields [23, 65-70]; at ultralow temperatures [71]; in the field of reverse-biased p−n junctions [30, 72-76]; and so on.

Nonetheless, the question of exciton condensation is still far from settled: much remains unclear concerning the diffusion of excitons and electron−hole drops [74, 77, 78] and the re-combination kinetics in the two-phase system exciton gas−electron-hole drop [28, 44, 46, 49, 79-81]; the kinetics of formation of liquid-phase nuclei and drop growth are problems that have hardly been studied at all. Intensive research is only just beginning on the condensation of excitons in wide-band semiconductors [82, 83] and on the interaction of ultrasound with EHD [84]. Next in line are investigations in superstrong magnetic fields, and assessement of the potential and practical utilization of the investigated phenomena.

The present paper is devoted to some of these little-studied questions. In the experi-ments, exciton condensation was studied by the light-scattering method, which permits direct measurement of EHD size and which is probably the only technique presently available for measuring exciton densities in crystals. By investigating how the EHD size and number density depend on experimental parameters like temperature, excitation intensity, and method of exci-tation, information can be obtained on the kinetics of formation of liquid-phase nuclei and on how they grow up to a steady-state size. Having such information, we can learn to control the process whereby nuclei are formed and thereby introduce into a crystal a specified number density of drops of prescribed size, an ability which clearly is of considerable importance as far as future practical applications are concerned. Our measurements were made on germanium, as it is in this semiconductor that EHD have been studied most thoroughly and good possibi-lities therefore exist for comparing experimental results obtained by different methods.

At the time that the present work was commenced, the well-known discussion on the na-ture of the "drop" line of germanium was still going on. A number of workers ascribed this line to emission accompanying the annihilation of excitonic molecules [85].[†] As the cross

---

[†] This point of view is given a basis in [85], where appropriate literature references are cited.

section for the scattering of light by a biexciton is negligibly small, light-scattering experiments were capable of resolving the matter in favor of the one model or the other. Light scattering by droplets of condensed phase was detected by Pokrovskii and Svistunova [52].

The present work consists of four chapters. The first reviews the literature on the problem of exciton condensation. An aim of this review is to bring out the basic properties of a two-phase exciton-gas—EHD system; for this reason we discuss only a restricted number of papers, sufficient to enable these properties to be described with adequate completeness. We note that a number of reviews have already been published on this subject [86-89].

In the second chapter, we consider methods of measuring EHD size. It is shown that all these methods, with the exception of the light-scattering method, are indirect, and yield only an approximate estimate of the size of the drops. The theory underlying the light-scattering method is presented and its application to EHD is discussed. We cite the results of measurements of drop size and number density and the density of electron—hole pairs in the drops. The experimental setup used to obtain these results is described.

In the third chapter we describe a high-sensitivity setup which we developed for the measurement of light scattering by EHD with simultaneous recording of the photoluminescence spectrum of germanium. The high sensitivity is attained by using a laser amplifier to amplify the light scattered by the drops and a helium—neon laser operating at 1.52 μm wavelength as the source of volume excitation. The method of preparation of the samples is described. The results of measurements of the dependence of drop size on temperature are given.

The fourth chapter is concerned with the kinetics of exciton condensation in germanium. A model of exciton condensation is discussed in which liquid-phase nuclei are produced through fluctuations of exciton density and in which the nuclei grow as a result of the diffusion of excitons towards their bounding surface. We cite the results of measurements of the dependence of drop size and number density on temperature, excitation intensity, and slope of the front of the exciting light pulse. The results are discussed on the basis of the above model of exciton condensation. An experiment on the determination of the coefficient of the surface tension of an electron—hole liquid is described and the results given. The chapter concludes by summarizing the main results of the work.

CHAPTER I

CONDENSATION OF EXCITONS AND THE PROPERTIES OF ELECTRON—HOLE DROPS

The Coulomb attraction between an electron and a hole in a semiconducting crystal leads, as is well known, to the formation of a coupled state of these two particles, called an exciton. Excitons can move around in the crystal quite freely, and as far as their internal structure is concerned they resemble, in many ways, the hydrogen atom or positronium. However, as the Coulomb interaction between an electron and a hole in a crystal is weakened by the presence of a dielectric constant, and as the effective masses of the electrons and holes are generally less than the free-electron mass, the binding energy of the particles in the exciton is of the order of a few millielectron volts and the Bohr radii of excitons in semiconductors is $\sim 10^{-6}$ cm. Excitons of this sort are usually called large-radius excitons, or Wannier—Mott excitons.

The large value of the Bohr radius of the exciton means that, even for an exciton density in the crystal as low as $\sim 10^{15}$-10^{16} cm^{-3}, the distance between excitons will be of the order of their Bohr radius and the exciton−exciton interaction energy will be of the order of the binding energy of the particles constituting the exciton. Under these conditions, collectivization of all electrons and holes is inevitable [1-3]. Let us consider the behavior of a system of excitons as their density is gradually increased for the case when the mean kinetic energy of the excitons is small compared with the binding energy of the particles constituting the exciton, i.e., we consider the temperature to be sufficiently low [3]. If the density of the exciton gas is small, the interaction between excitons is negligible and the system can be regarded as an elemental ideal gas. As the exciton density increases, excitonic molecules (biexcitons) can be produced. However, as the effective electron and hole masses in semiconductors are approximately the same, it follows that the amplitude of the zero-point oscillations in a biexciton is of the order of the Bohr radius of an exciton and that the energy of these oscillations is of the order of the exciton binding energy. The dissociation energy of an excitonic molecule must therefore be considerably less than the binding energy of the particles constituting the exciton. Calculations show that in Ge the dissociation energy of a biexciton is less than 0.1 eV [90, 91]; the experimentally determined binding energy of the electron and hole constituting an exciton amounts, however, to 3.8-4.15 meV [92-94]. An analogous situation probably also obtains in silicon. The small dissociation energy of the excitonic molecule means that a phase transition in a system of excitons must be similar, in many ways, to the condensation of vapors of the alkali metals to the liquid phase, i.e., when the mean exciton density in the sample reaches a certain definite (temperature-dependent) value, regions of metallic electron−hole liquid must be formed, the density of electrons and holes in which greatly exceeds the mean carrier density in the sample. In this manner, attainment of the threshold conditions results in the formation in the crystal of electron−hole drops "immersed" in an exciton gas. The exciton density above the liquid drops differs from the thermodynamic equilibrium value [3]:

$$n_{\text{от}} = \nu \left(\frac{M^* kT}{2\pi\hbar^2} \right)^{3/2} e^{-\varepsilon_0/kT}, \tag{1.1}$$

since the electrons and holes from which the drops and the excitons are composed have a finite lifetime and continuous excitation is required to maintain a given density. In (1.1) ν is the multiplicity of the degeneracy of the ground exciton state, M^* is the effective exciton density-of-states mass, ε_0 is the binding energy per pair of particles in the liquid phase measured from the ground exciton level, and T is the absolute temperature.

We mention at this point that there is a considerable difference between an electron−hole liquid and an ordinary electron−hole plasma. In a plasma, the kinetic energy of the electrons and holes is large compared with their energy of interaction; accordingly, if a plasma cloud is formed in some manner in some region of a semiconductor, it begins to disperse by diffusion throughout the crystal. In contrast to this, the particles of an electron−hole liquid are held together by the forces of interaction and the liquid occupies a limited volume, its density being the same everywhere in this volume.

Parameters characterizing the basic properties of the condensed phase are the binding energy ε_0 and the carrier density in the liquid phase n_0. These quantities were calculated in [19] for Ge and in [21] for Si. The model used in these papers to describe the liquid phase is as follows: electrons and holes in the liquid are degenerate; the energy of the particles is additively compounded of an electron part and a hole part. The mean energy per pair of particles in the liquid is then

$$\tfrac{3}{5}(E_e + E_h) + E_x^{ee} + E_x^{hh} + E_c^{ee} + E_c^{hh}, \tag{1.2}$$

where E_e and E_h are the Fermi energies for electrons and holes respectively; E_x^{ee} and E_x^{hh} are the exchange energies for electrons and holes; and E_c^{ee} and E_c^{hh} are the correlation energies for electrons and holes. Since E_c, $E_h \sim n_0^{2/3}$, and E_x^{ee}, $E_x^{hh} \sim n_0^{1/3}$ and are negative (the quantities E_c^{ee} and E_c^{hh} are also negative, although they depend more weakly on n_0 than do E_x^{ee} and E_x^{hh}), it follows that expression (1.2) has a minimum when the quantity n_0 equals the equilibrium density of the carriers in the liquid. It was shown in [19, 21] in this manner that $n_0 = 2 \cdot 10^{17}$ cm^{-3} for germanium and $n_0 = 3 \cdot 10^{18}$ cm^{-3} for silicon; the calculations utilized the expressions for the Fermi and exchange energies familiar from the theory of metals, and the correlation energy was evaluated using Wigner's formula [95].

More rigorous calculations of the quantities n_0 and ε_0 have, in fact, been carried out to date, but their results do not differ appreciably from those of [19, 21]. Thus, in [91, 96] the values found for germanium are $n_0 = 1.8 \cdot 10^{17}$ cm^{-3}, $\varepsilon_0 = 1.7$ meV; in [91] the values found for silicon are $n_0 = 3.4 \cdot 10^{18}$ cm^{-3}, $\varepsilon_0 = 5.7$ meV. In [9] the values found for germanium are $n_0 = 2 \cdot 10^{17}$ cm^{-3} and $\varepsilon_0 = 2.5$ meV and for silicon are $n_0 = 3.1 \cdot 10^{18}$ cm^{-3} and $\varepsilon_0 = 6.3$ meV.†

The particle density in the liquid phase has been determined experimentally in various ways: from the ratio of the free-exciton (FE) and EHD radiative lifetimes [28]; from the ratio of the integrated intensities of the TA- and LA-phonon-assisted EHD recombination lines [28]; from magnetooptical measurements [67–70]; and by the light scattering method [53]. The best results probably come, however, from measurements of the EHD absorption of long-wavelength IR radiation [6, 39, 40] and from analysis of the EHD recombination line shape [19, 21, 24, 29, 34, 37, 38].

In a study of the absorption of germanium in the far infrared for excitation of nonequilibrium carriers by light from an incandescent lamp, Vavilov et al. [6] discovered, at temperatures below 2°K, a broad absorption band peaking at around 9 meV. Long-wavelength IR emission was also observed, the peaks of the emission and absorption bands coinciding. These effects were investigated in detail in [39, 40]. Plots of the absorption and emission signals vs. excitation intensity were of a threshold character and were explained by the absorption and emission of light by droplets of condensed phase. If an electron–hole drop is situated in the field of an electromagnetic wave and its dimensions are much less than the wavelength, then it behaves as a dipole in a uniform alternating field. The cross section for absorption of electromagnetic radiation then has a clearly expressed resonance at a frequency $\omega_0 = \omega_p / \sqrt{3}$, where $\omega_P = (4\pi e^2 n_0 / \varkappa_0 m^*)^{1/2}$ is the plasma frequency [$\varkappa_0$ is the dielectric constant of the crystal, m* is the reduced electron (hole) effective mass]. The value of n_0 found from the resonance frequency was $n_0 = 2 \cdot 10^{17}$ cm^{-3}. The collision frequency was also estimated from these measurements, and was found to be $\sim 9 \cdot 10^{12}$ sec^{-1}, which determines the damping time of the plasma oscillations.

The density of the electron–hole liquid is usually found from an analysis of the EHD recombination line shape. An advantage of this method is that the binding energy ε_0 can also be determined from spectroscopic measurements. The line due to EHD recombination radiation in germanium was detected in [5]. The peak of this line is shifted by 4.6 meV towards longer wavelengths away from the peak of the FE line. The appearance of the EHD recombination line was of a threshold character as regards variation of temperature and excitation intensity. EHD recombination radiation in silicon was first observed by Haynes [98], although he incorrectly interpreted this radiation as arising in the decay of excitonic molecules. The model proposed in [98] for the decay of a biexciton was based on an obsolete value of the exciton binding energy in silicon (8 meV). At the present time it is reliably established that the

---

† The results of calculations of n_0 and ε_0 carried out by various authors are summarized in [97].

value is 14 meV [99], with the result that the model of biexciton recombination considered in [98] is in internal contradiction. The careful investigations reported in [20, 21, 27] showed that the emission observed in [98] is due to the recombination of the electrons and holes constituting electron—hole drops.

The method of analysis of the EHD luminescence line shape is used in [19, 21, 24]. The EHD spectral radiation density is found on the assumption that the probability of radiative transitions is independent of the electron (hole) energy. The spectral density was found using the normal expression employed for indirect semiconductors with a quadratic dispersion law, and expression (1.2) for the energy of the particles in the liquid phase. The calculated EHD emission spectra were compared with the experimental spectra, there being only a single adjustable parameter, n_0. The particle density of the liquid found from a best-fit agreement of the experimental and computed spectra was found to be: for germanium $n_0 = 2.6 \cdot 10^{17}$ cm^{-3} [19]; for silicon $n_0 = 3.7 \cdot 10^{18}$ cm^{-3} [21].

Analysis of the EHD recombination line shape has since been used many times for germanium. In [34] the lineshape was investigated at various temperatures. These measurements showed that the electron (hole) Fermi energy varies in accordance with the theory of metals, the entropy of the electron gas in which is directly proportional to the absolute temperature. The compressibility of the electron—hole liquid measured in [34] was found to be $3.7 \cdot 10^{-18}$ cm^3/meV. We note that the compressibility of the electron gas in metals has still not been determined experimentally. The dependence of the EHD luminescence line shape in germanium on excitation intensity was measured in [26, 38]. If the density of nonequilibrium carriers was less than $\sim 10^{17}$ cm^{-3}, the line shape did not change as the excitation energy was varied. At greater densities of nonequilibrium carriers, the EHD recombination line broadened and its peak was shifted towards shorter wavelengths: the electron—hole liquid filled the whole sample and, with further increase of the excitation intensity, its density increased and the binding energy decreased.

Precision measurements of the EHD and FE luminescence line shapes at various temperatures were made in [37]. The measurements were made on samples of ultrapure germanium, and the excitation intensity at each temperature was such that the FE and EHD luminescence lines were of approximately equal intensity (Fig. 1). This was done to achieve best resolution of the FE and EHD lines. The FE lineshape was analyzed on the basis of Elliot's theory [100]. The error introduced by the finite width of the spectrometer slit was taken into account in the

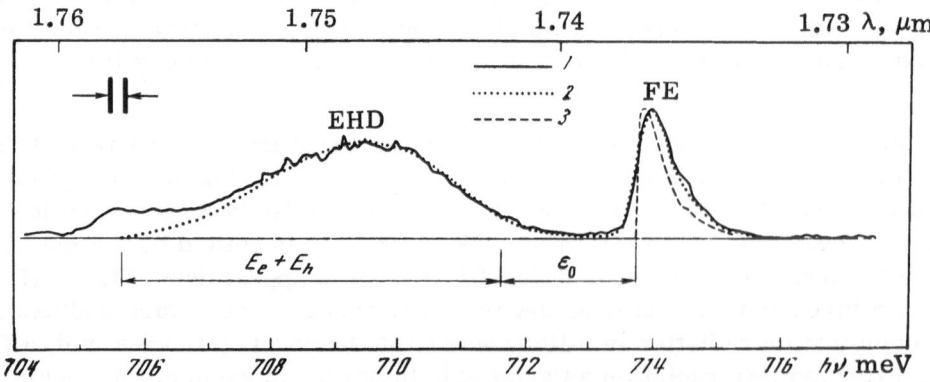

Fig. 1. Recording of spectrum and result of theoretical analysis of EHD and FE luminescence line shapes in Ge at 3.04°K [37]. (1) Experiment; (2) theory, allowing for finite width of spectrometer slit; (3) ignoring finite width of slit.

analysis of the spectra. It was shown that the electron—hole density in the liquid phase depends on temperature according to the law $n_0 = 2.38 \cdot 10^{17}(1 - 0.01\ T^2)$ cm^{-3}. The value found for the binding energy was $\varepsilon_0 = 2.06 \pm 0.15$ meV.

Let us consider for a moment the determination of n_0 from magnetooptical measurements. The EHD recombination radiation from Ge in strong magnetic fields was investigated in [23, 65, 66]. In fields from 40 to 70 kOe the EHD line was found to split into two lines, which is connected with the Landau quantization of the energy spectrum of the degenerate carriers in the electron—hole liquid. In fields greater than 70 kOe, the splitting between the Landau subbands becomes greater than the hole Fermi energy, and a single line remains in the spectrum. Oscillations of the integrated intensity of the EHD luminescence were observed in [67], and were explained by the passage of the Landau levels through the EHD-electron Fermi level. The period of the oscillations of the integrated intensity plotted against the reciprocal of the field was used to determine the liquid-phase density $n_0 \simeq 2 \cdot 10^{17}$ cm^{-3}. Quantum oscillations of the intensity of EHD recombination radiation from silicon have also been observed [70]. Measurements made in strong magnetic fields in germanium disclosed oscillations of the absorption coefficient in the far IR [68] and oscillations of the intensity of the long-wavelength IR EHD luminescence [69]. A theory of the electron—hole liquid in semiconductors in a magnetic field was proposed in [101].

The binding energy of the particles in an electron—hole liquid has been determined not only from spectroscopic measurements but also from the temperature dependence of the intensity of the FE and EHD lines and from the temperature dependence of the threshold excitation intensity. Under steady-state conditions, the flux of excitons to the surface of the drop must equal the rate at which electrons and holes recombine within it. If the dimensions of the drop are small compared with the exciton mean free path, then [19]

$$^{4}/_{3}\,\pi R^3 n_0/\tau_0 = 4\pi R^2 (n - n_{\mathrm{or}})v_{\mathrm{T}}, \tag{1.3}$$

where R is the radius of a spherical drop, τ_0 is the carrier lifetime in the liquid phase, and v_{T} is the exciton thermal velocity. It can be shown using (1.3) that, provided the temperature is not too low, the FE luminescence intensity is given by [21]

$$I_{\mathrm{FE}} \sim \exp\left[-\frac{\varepsilon_0}{k}\left(\frac{1}{T} - \frac{1}{T_0}\right)\right]. \tag{1.4}$$

and that at high temperatures the EHD luminescence intensity is given by

$$I_{\mathrm{EHD}} \sim \left\{1 - \exp\left[-\frac{\varepsilon_0}{k}\left(\frac{1}{T} - \frac{1}{T_0}\right)\right]\right\}^3. \tag{1.5}$$

Here T_0 is the threshold temperature for a given rate of generation. Relationships (1.4) and (1.5) can be checked experimentally; the only adjustable parameter in them is ε_0. For germanium, using (1.5), the value of ε_0 is found to be 1.5 meV [86]. The temperature dependence of the FE and EHD luminescence intensities gives, for silicon, $\varepsilon_0 = 2$ meV [21, 86]. We note, however, that Eqs. (1.4) and (1.5) are derived on the assumption that the EHD number density, the number of drops per unit volume, is independent of temperature. It will be shown in Chapters III and IV that this assumption is probably not justified.

In accordance with (1.1), the binding energy can be determined from the temperature dependence of the threshold excitation intensity. Such measurements have been made several times for Ge. In [19] the threshold was fixed via equality of the intensities of the FE and EHD lines, and a value of 2.7 meV was found for ε_0. In [30, 31] the threshold was fixed via the appearance of EHD luminescence and current pulses through a reverse-biased p—n junction; in

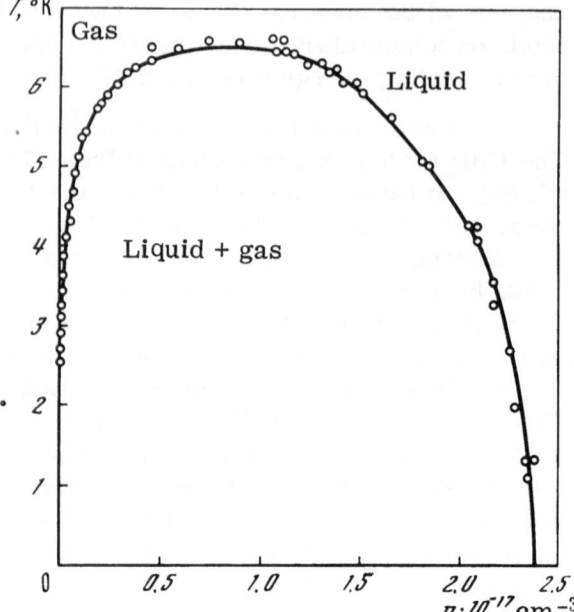

Fig. 2. Phase diagram of FE−EHD system [35, 36].

[45, 46] it was fixed via the appearance of free-carrier cyclotron resonance. The values obtained for ε_0 were 1.54 meV [30] and 1.38 meV [46].†

The temperature dependence of n_{OT} given by (1.1) constitutes one of the branches of the phase diagram of the system: exciton gas−electron-hole liquid. This branch was constructed in [30]. Another branch is given by the curve depicting the temperature dependence of the equilibrium density of the liquid [38]. The phase diagram for Ge has been constructed via luminescence data [35, 36] (Fig. 2). For Si, part of the liquidus line near the critical point was constructed in [102].

It should be noted that the binding energy ε_0 found from temperature measurements is generally less than ε_0 determined from spectroscopic data. Perhaps there is nothing surprising about this, as the thermal energy of activation is usually less than the optical. However, in the present case, it must be remembered that the exciton gas −EHD system is a nonequilibrium one, so that thermodynamic relationship (1.1) may not be fulfilled. It is shown in [103] (this question is also considered in Chapter IV) that the deviation of the temperature dependence of the threshold generation rate from the dependence predicted by (1.1) should be at its greatest at low temperatures. This effect has been observed experimentally at temperatures ~2°K [31, 46]. Furthermore, temperature dependence of the rate at which nuclei of liquid phase are formed may distort the results of temperature measurements of the binding energy.

An interesting phenomenon, termed optical hysteresis by the authors, was observed in [30]. For a given excitation intensity, the volume of liquid phase in the sample depended on how this excitation intensity was reached. If a given excitation intensity was reached by steadily increasing the pump power, then the volume of liquid was less than when this same intensity was reached by decreasing the power. This effect is probably connected with a dependence of the rate of formation of liquid-phase nuclei on the degree of supersaturation of the exciton vapor.

---

† The results of measurements of ε_0 by various methods are compared in [31].

The rate at which liquid-phase nuclei are formed must be greatly increased if condensation centers are present in the sample. Defects in the crystal structure can serve as condensation centers. The role played by impurities in the phenomenon of exciton condensation has been investigated in a number of papers. It was shown in [22, 26] that increasing the concentration of shallow impurities in a Ge crystal increases the threshold temperature for a given excitation intensity. This is connected with a reduction of the effective volume in which condensation takes place, and probably also with an impurity-enhanced increase in the rate of formation of nuclei. The no-phonon luminescence line of the condensed phase was observed in [26, 32] at an impurity concentration of 10^{16} cm^{-3}; a reduction of the carrier lifetime in the liquid phase was also observed. Optical hysteresis in impurity Ge was observed in [104], and it was shown that this effect is insignificant in the presence of impurities.

Interesting results have been obtained from studies of Si doped with boron [20, 21, 24]. At small excitation levels, the recombination spectrum contains lines from free excitons and excitons bound to boron atoms. As the excitation level is increased, additional lines shifted towards longer wavelengths away from the bound-exciton line appear in the spectrum. These lines were ascribed to the formation on boron atoms of complexes consisting of more than one exciton. As the excitation intensity is further increased, a long-wavelength tail appears in the spectrum which gradually develops into a broad peak corresponding to the luminescence line of the electron−hole liquid. Bound multiexciton complexes are being widely investigated at the present time. A beautiful picture was obtained in a study of Si doped with lithium, the spectrum of the multiexciton complexes in which consists of ten lines [105]. These results indicate that bound multiexciton complexes serve as condensation centers in doped silicon. A similar situation probably also obtains in germanium [33, 106].[†]

We dwell for a moment on results obtained on the kinetics of formation of electron−hole drops. If liquid-phase nuclei that are capable of growth have appeared in a crystal, then they grow to dimensions that are determined by the exciton density and by the temperature of the sample. The growth time of electron−hole drops has been measured via the delay in the appearance of current pulses through a reverse-biased p−n junction [74] and via the buildup of EHD recombination luminescence [44, 56] for excitation of Ge by short light pulses; the value found was 0.1-0.7 μsec. The growth time was found to decrease with increasing excitation in intensity [74]. Some conclusions on the kinetics of condensation can be drawn from the dependence of the EHD luminescence intensity on excitation level. It can be seen from (1.3) that, for $n \gg n_{OT}$ and for a rate of recombination in the gas phase much greater than in the liquid phase, we have, for the intensity of the EHD line,

$$I_{EHD} \sim g^3, \tag{1.6}$$

if the drop number density in the crystal does not depend on the generation rate g. If, on the other hand, the nonequilibrium carriers recombine primarily in the liquid phase, we have, for the volume of the latter, and so also I_{EHD}, the following:

$$I_{EHD} \sim g. \tag{1.7}$$

Photoluminescence investigations [19, 22, 86, 108] and long-wavelength IR absorption and emission measurements [6, 39, 40] performed under steady-state excitation conditions using samples of pure Ge showed that Eq. (1.6) and (1.7) correctly describe the experimental results. Similar results have also been obtained in silicon [20, 21, 24, 86]. However, the

---

[†] Papers on multiexciton complexes are reviewed in [107].

strong dependence of the EHD number density on excitation level (see Chapter IV) suggests rather strongly that the interpretation of these experiments with the aid of relationship (1.3) is not the only one possible. In doped germanium, the volume of the liquid phase near the condensation threshold is observed to grow very rapidly with increasing excitation level [22]. Experiments using pulsed excitation show that the dependence of the intensity of the recombination radiation on generation level is quadratic [80, 109].

Recombination kinetics in an exciton gas—EHD system have been studied in many papers. Lifetimes have been determined in a number of ways by measuring the following: the frequency dependence of the intensity of the recombination radiation [19]; the decay of the recombination radiation [26, 28, 56, 79, 80, 81, 110] and of cyclotron resonance signals [46, 49] for pulsed excitation conditions; and the diffusion coefficient by using the moving glowing spot technique [24, 77]. The lifetime of carriers in the liquid phase was found to be 20-40 μsec and the FE lifetime 2-8 μsec. According to the results of [19, 28, 111], the quantum yield of radiation from the condensed phase in Ge amounts to 25-80%, whereas for excitons this quantity lies in the range 1-10%. In silicon, due to the large value of the Auger-recombination coefficient, the EHD-luminescence quantum yield is ~0.05% and the lifetime of carriers in the liquid phase is ~0.2 μsec. The FE lifetime in Si is almost an order of magnitude greater than this [86].

Studies of the recombination kinetics in an FE—EHD system at various temperatures have shown that in Ge at temperatures below ~2.5°K, when thermal ejection of carriers from the electron—hole drops can be neglected, recombination goes primarily through the liquid phase. At higher temperatures, evaporation of carriers from the EHD surface becomes important and recombination of the carriers bound into excitons becomes the dominant process. The time for which the liquid phase exists is then reduced, since under these conditions it is determined by the rate of evaporation and by the FE lifetime [28, 45, 46, 56, 79-81, 110]. It should be noted that qualitative relationships describing the recombination kinetics in the given nonequilibrium gas—liquid system are not readily compared directly with experiment as they contain several adjustable parameters.

Experiments on the recombination kinetics in an exciton gas—EHD system resulted in the discovery of a new type of microwave breakdown in germanium [43, 47, 48]. Exciton breakdown was induced by a microwave pulsed only if the latter was shifted in time relative to the exciting light pulse. Furthermore, the time interval between the exciting and microwave pulses increased with decreasing amplitude of the microwave pulse. This effect was explained by assuming that electron—hole drops are efficient centers for the capture of free carriers [47, 48]. The capture rate is then proportional to the total surface area of the drops. Accordingly, breakdown occurred only when the dimensions of the drops were reduced to such an extent in the recombination process that the multiplication rate of free carriers due to impact ionization of excitons started to exceed the capture rate.

One of the most striking properties of EHD is their high mobility in nonuniform fields [3]. Due to the Fermi energy distribution of the carriers in an EHD, the momentum relaxation time of the drop as a whole is very large, amounting to ~10^{-8} sec. Motion of EHD in an inhomogeneous strain field was reported in [7]. Inhomogeneous straining of a Ge sample produced a strong reduction of the intensity of the EHD line, an effect which was explained in [7] by the motion of the drops towards the region of greatest strain. It was shown that for uniaxial straining at pressures less than a certain value P_{cr}, the binding energy per particle pair in the drop and the equilibrium electron—hole density of the drop are reduced. This effect is explained by the relative motion of the conduction band minima of germanium and by the redistribution of the electrons among the minima. The redistribution of the electrons ceases at pressures greater than P_{cr}, and with further increase of the compression the binding energy does not change.

These investigations were continued in [23, 26, 59]. Measurements were performed for various directions of uniaxial compression. When compression was applied in the [100] direction, all conduction-band minima remained equivalent; the reduction of the binding energy at pressures less than P_{cr} is explained by splitting of the valence band. Under conditions of inhomogeneous straining, the drops moved to the region of maximum strain and luminesced from this region. This was checked by experiments on samples of special shape [61], when it was possible to calculate the mechanical stress in the region of maximum strain. A direct calculation showed that the shift of the EHD line, which was used to determine the change in the binding energy, corresponded to the pressure in the region of maximum strain. In [56, 112] EHD motion was monitored via the absorption of 3.39 μm IR radiation. These direct measurements showed that, in an inhomogeneous strain field, the drops manage to move a distance of a few mm during their lifetime. The EHD velocity of motion was determined with the aid of pulse measurements; it was found to be $\simeq 3 \cdot 10^5$ cm/sec [56]. Heating of electron−hole drops due to interaction with acoustic phonons during the motion of the drops in an inhomogeneous strain field was reported in [63]. At a temperature of 4.2°K, electron−hole drops in Ge were heated up to $\simeq 6$°K, the rise in temperature of the drops being proportional to the pressure gradient.

Studies of the polarization of the EHD recombination luminescence in uniaxially strained Si [62] and Ge [56, 64] showed that a significant degree of polarization is observed only at pressures greater than P_{cr}, as is to be expected for a Fermi energy distribution of the carriers in the drops.

The EHD diffusion coefficient can be determined from the motion of the droplets in an inhomogeneous strain field. The time taken by a drop to travel a given distance has been directly measured in experiments using samples of special shape for which the strain gradient can be calculated [56]. The velocity of the drops can thus be found. The shift of the EHD line was used to determine the force acting on an electron−hole pair in the drop, which in turn enables the drop momentum relaxation time to be calculated. By prescribing the drop dimensions that are typical for the given experimental conditions and which are frequency determined by the light-scattering method [52-58], it is possible to estimate the mobility of the drops and, using the Einstein relationship, their diffusion coefficient. Such an estimate gives a value of $\sim 10^{-3}$ cm^2/sec for the diffusion coefficient. This method, and the method used in [113, 114] to determine the charge on electron−hole drops, are probably the most direct methods of determining the EHD diffusion coefficient available at the present time.

The difference in the levels of the electron and hole chemical potentials in an electron−hole liquid means that electrons and holes must evaporate from the surface of the drop at different rates until such time as the electrochemical potentials of the two types of carrier become comparable; the drop must then acquire a certain electrical charge. Calculations made in [11] show that the drops have a negative charge of value $\sim 100e$. Calculations carrier out by other authors lead, however, to a positive charge [14].

Pokrovskii and Svistunova [113] carried out the following experiment to determine the charge on an electron−hole drop: the spatial distribution of the EHD luminescence was measured in an electric field and without it. The direction in which the spatial distribution is displaced in the electric field gives the sign of the EHD charge; the magnitude of the displacement enables the drop mobility to be calculated, as the magnitude of the displacement of the EHD luminescence distribution is approximately equal to the distance traveled by the charged drops in the electric field in a time equal to their lifetime. The mobility found in these experiments approximately coincides with the value found from the motion of drops in an inhomogeneous strain field [56] if we take the drop charge to be $\sim 100e$. Also, the sign of the charge was found to be negative, in agreement with [11].

There is an interesting possibility whereby the sign of the drop charge can be changed. If the sample is compressed in the [111] direction, all the electrons go over into the one conduction-band minimum at a pressure of $\simeq 200$ kg/cm^2; the electron Fermi energy then exceeds the hole Fermi energy, and the drop acquires a positive charge. An experiment of this sort was carried out by Pokrovskii and Svistunova in [114].

The EHD "diffusion" coefficient is determined in most papers by measuring the dimensions of the region occupied by the drops [24, 49, 74, 77, 78, 86, 114]. There is, however, a colossal variation (of around four orders of magnitude) in the results of different measurements. In [24, 49, 77, 86] where a relatively small excitation intensity was used, the diffusion coefficient $D \lesssim 0.1$-1 cm^2/sec. This value exceeds by at least two orders of magnitude the EHD diffusion coefficient found from experiments on the motion of drops in an inhomogeneous strain field [56]. A value of $D \simeq 150$ cm^2/sec is reported in [74]; in [78] the diffusion coefficient $D = 25$-500 cm^2/sec depending on the experimental conditions, and D was observed to increase with increasing excitation level. It is shown in [114] that the "diffusion" coefficient increases from 0.8 to 80 cm^2/sec as the excitation level is increased from 14 to 600 mW. The results of these experiments suggest that some sort of mechanism of EHD drift motion exists, the drift velocity increasing with increasing excitation intensity. The idea that the drops are carried along by excitons when an exciton density gradient is present is put forward in [76, 114, 115]. Estimates show, however, that this effect must be very small, especially at low temperature, when the exciton density is insignificant; also, experiment gives a contradictory result [78]. The idea that the drops are carried along by the phonon wind arising in the thermalization of nonequilibrium carriers in the region of excitation is advanced in [116], and the results of an experimental observation of this effect are cited. The notion of a phonon wind completely explains the results of experiments on the spatial distribution of electron−hole drops [24, 49, 57, 74, 77, 78, 86, 114, 117].

An important parameter characterizing the properties of the two-phase FE − EHD system is the size of the drops. Far-IR absorption measurements [40] and microwave conductivity data [45, 49] show that the drops occupy 10^{-5}-10^{-3} of the volume containing the nonequilibrium carriers. The size of the drops has been estimated in a number of ways: from long-wavelength IR absorption data [6, 39, 40]; from large photocurrent pulses induced in a reverse-biased p−n junction [73, 75, 76]; from EHD recombination kinetics as studied by cyclotron resonance [46] and via recombination radiation [110]; from microwave breakdown of excitons [47, 48]. The size is found to lie in the range 1-10 μm. Direct measurement of drop size by the light-scattering method gives for the EHD radius a value of from 2 to 10 μm depending on temperature and excitation intensity [52-58].

The authors of a number of recent papers [50, 51, 81, 117] consider that they have observed large drops, of dimensions from 0.16 to 1.0 mm. The experiments on the observation of large drops were carried out on samples of ultrapure germanium with a residual impurity content of less than 10^{10} cm^{-3}. Feldman [117] measured the spatial distribution of the EHD luminescence for various excitation intensities. This distribution differed from that obtained in [78, 114] at large generation levels, and was in good agreement with the theoretical expression for a uniform spherical luminescing region. However, a light-scattering experiment performed under similar conditions showed that this region consists of small drops of radius ~2 μm [57]. We note that the geometrical shape of the region in which the electron−hole drops are concentrated accords well with the idea of EHD entrainment by nonequilibrium phonons [116].

Westervelt et al. [81], who were investigating EHD recombination kinetics, suggested that only a single drop is formed in the sample under the conditions of their experiment. They did not provide sufficient evidence in support of this suggestion; furthermore, the rate equations which they employed to analyze the results of their experiments ignored the diffusion of excitons towards the surface of the drops, which, for sufficiently large drops, can have a considerable effect on the interpretation of the results.

In [50, 51] large drops were observed by the method of dimensional resonance. More refined investigations carried out by these same authors [118] showed that the large drops in these experiments were observed on account of inhomogeneous straining of the sample. In special experiments with uniaxially strained Ge crystals it was shown that the large drops have a density that is almost an order of magnitude less than drops in an unstrained sample (and, consequently, a lifetime that is an order of magnitude greater). A large drop in a uniaxially strained crystal was photographed by Wolfe et al. [119].

The nonequilibrium carriers in the experiments described above were generated by optical sources of surface and volume excitation operating in the cw and pulse modes. The FE−EHD phase transition has also been investigated using electron-beam excitation [25] and with the aid of double injection in germanium [120, 212] and silicon [122]. Investigations of exciton condensation using double injection are particularly interesting, as they show that the processes responsible for the binding of nonequilibrium carriers into excitons and electron−hole drops may play an important role in the operation of cryoelectronic devices.

We conclude this brief review of the literature by noting that the gas−liquid transition in a system of excitons has, to date, been studied in reasonable detail only in germanium and silicon. Work on other semiconductors is only just beginning. Similar phenomena will probably also be observed in other semiconductors, although the nonequilibrium system is more complex [82, 83].

CHAPTER II

DETERMINATION OF SOME EHD PARAMETERS BY THE LIGHT-SCATTERING METHOD

An electron−hole liquid in a crystal has the form of individual drops immersed in an exciton gas. Accordingly, the size of the drops is one of the most important parameters characterizing the properties of this nonequilibrium system. Light scattering provides a direct method of measuring the dimensions of optical nonuniformities, and, since the refractive index of an electron−hole drop differs from that of the crystal, this method can be used to determine the dimensions of droplets of condensed phase. Simultaneous measurement of the absorption and scattering of light enables the electron−hole density of the liquid and the drop number density in the crystal to be determined.

In this chapter we consider various methods that have previously been used to determine EHD dimensions. It is shown that all these methods are indirect and very approximate. We discuss the Rayleigh−Gans theory of scattering, on the basis of which our experimental results are analyzed, and we consider the possibilities of the scattering method as applied to electron−hole drops. An experimental setup for investigating light scattering by EHD in Ge is described in detail, and the results of measurements made using the setup are discussed.

1. Methods of Measuring EHD Size

The dimensions of electron−hole drops were first estimated from long-wavelength IR absorption data [6, 39, 40]. The coefficient of absorption of long-wavelength IR radiation by EHD is proportional to the average (over the volume of the sample) density of electron−hole pairs in the liquid phase. Thus, by measuring the absolute value of the absorption coefficient, the average density of condensed carriers can be found. If, also, we graph the dependence of the absorption coefficient on excitation level, then, by (1.6) and (1.7), this graph should consist of two parts, a linear part and a cubic part, the point of inflection corresponding to equal rates of recombination in the liquid and gaseous phases. By equating the rates of carrier recombination in drops and excitons and utilizing relationship (1.3), we have for the radius of the drops at the point of inflection:

$$R = \frac{3v_\tau \tau}{n_0} N_\Sigma, \tag{2.1}$$

where N_Σ is the average (over the volume of the sample) carrier density in the liquid phase at the point of inflexion, and τ is the exciton lifetime.

An estimate of the EHD radius using (2.1) was made in [39, 40]; the result obtained was $R \simeq 4 \ \mu m$. The disadvantages of this method of measuring drop radius include the following: a certain arbitrariness in the determination of the point of inflection; difficulties in measuring the absolute value of the absorption coefficient; the need to measure the exciton lifetime. Also, this method can be used to estimate the EHD radius only at the point of inflection, and for a more exact estimate relationship (2.1) must be replaced by a more general expression which allows for the diffusion of excitons towards the surface of the drops.

An interesting method of determining the dimensions of EHD was used in [73, 75, 76]. In these papers the number of particles in an electron—hole drop was measured via the amplitude of the large photocurrent pulses induced in a reverse-biased p—n junction when exciton condensation occurs near it. In [75, 76] an attempt was made to find the EHD size distribution. It was shown that the size distribution is strongly dependent on the distance between the p—n junction and the region of excitation; the greater this distance, the more uniform the size distribution became. Furthermore, if the p—n junction was prepared from p-type germanium, the sizes of the drops were almost the same; n-type samples, on the other hand, gave a more diffuse distribution. In other words, the drops enter the field of the junction at different stages of their growth. In this case, the size distribution must depend on the structure of the junction used.

The size of electron—hole drops has recently been estimated from experiments on drop recombination kinetics [81, 110]. In these experiments the recombination kinetics were investigated via the decay of the intensity of the EHD luminescence line for pulsed excitation. Solving the appropriate rate equations leads to quite a simple relationship between EHD size and the total carrier recombination lifetime in the liquid phase, subject to the assumption that the flux of excitons onto the drops can be neglected compared with the ejection of excitons from the drops. Size determination by this method requires knowledge of the FE work function. Also, it is difficult to precisely fix the total decay time of the luminescence signal, while the calculation of this time involves solving the rate equations with adjustment of a number of parameters and a computer calculation. It should be noted that exact rate equations should allow for the diffusion of excitons towards the surface of the electron—hole drops. It was probably a failure to allow for this that led to the rapid increase in the size of the drops at temperatures ~ 4°K in [110].

The recombination kinetics of the carriers bound into electron—hole drops was investigated in [46] through the decay of the cyclotron resonance signal. The decay time in this paper was fixed in terms of a definite preselected signal level which corresponded to the noise level. The size of the drops estimated from these measurements was ~8 μm.

The size of electron—hole drops has also been determined from experiments on the microwave breakdown of excitons [47, 48]. In these papers the electron—hole drops are regarded as effective trapping centers for free electrons, as a result of which they delay the development of an electron avalanche in exciton breakdown. The rate at which electrons are captured by drops is

$$T_e = 4\pi R^2 N v_\varepsilon = 3N_\Sigma v_\varepsilon/n_0 R, \tag{2.2}$$

where v_ε is the average random electron velocity in the microwave field of given intensity, and N is the drop number density. It can be seen from (2.2) that the EHD radius R can be found by measuring Γ_e and N_Σ. The average (over sample volume) density of carriers bound into EHD at the moment of breakdown was estimated from the energy of the exciting laser pulse allowing for the carrier recombination rate in the liquid phase. The rate at which electrons are captured by the drops was measured in two ways: via the breakdown damping time constant, and via the dependence of the breakdown threshold on the duration of the microwave pulse. These measurements gave: $\Gamma_e \approx 3 \cdot 10^6$ sec^{-1} and R $\sim$ 10 μm.

In this manner, it is apparent from this brief review of methods of measuring the size of electron-hole drops that all of these methods require knowledge of a greater or lesser number of additional parameters, some of which can be estimated only very approximately.

2. The Rayleigh — Gans Theory of Scattering

We consider the scattering of a plane-polarized electromagnetic wave propagating along the z axis by a spherical particle of radius R. We shall find the intensity of the light scattered by the particle at an angle θ to the z axis in the plane in which the incident and scattered rays lie (this plane is called the scattering plane). In the case of a spherical particle, we have for the electric vector of the scattered wave [123]:

$$E_\perp = S_1(\theta) \frac{\exp[-ik(r-z)]}{ikr} E_0 \cos \varphi, \tag{2.3}$$

$$E_\parallel = S_2(\theta) \frac{\exp[-ik(r-z)]}{ikr} E_0 \sin \varphi. \tag{2.4}$$

Here $E_\perp$ and $E_\parallel$ are respectively the projections of the electric vector of the scattered wave upon the normal to the scattering plane and upon the scattering plane itself, $S_1(\theta)$ and $S_2(\theta)$ are amplitude functions, $k = 2\pi m_0/\lambda_0$ is the magnitude of the wave vector in the medium containing the scattering particle, m_0 is the refractive index of the medium, λ_0 is the free-space wavelength, r is the distance from the scattering particle to the point of observation, E_0 is the magnitude of the electric vector of the incident light wave, and φ is the angle between E_0 and the normal to the scattering plane. The intensity of the light scattered at an angle θ in the scattering plane will be

$$I_p = E_\perp^2 + E_\parallel^2. \tag{2.5}$$

Substituting Eqs. (2.3) and (2.4) in Eq. (2.5), we obtain

$$I_p = [|S_1(\theta)|^2 \cos^2 \varphi + |S_2(\theta)|^2 \sin^2 \varphi] I_0/k^2 r^2, \tag{2.6}$$

where $I_0 = E_0^2$ is the intensity of the electromagnetic wave incident upon the particle.

It can be seen from (2.6) that, in order to find the intensity of the scattered light, we have to calculate the amplitude functions $S_1(\theta)$ and $S_2(\theta)$. These functions can be evaluated quite easily using the Rayleigh—Gans theory provided $|m-1|$ and $2kR|m-1| \ll 1$, where $m = (m_{Re} - im_{Im})/m_0$ is the complex refractive index of the scattering particle with respect to the medium. The first of these conditions means that, when calculating the alternating dipole moment of the particle, we can regard the electric field of the light wave within the particle as the same as in the medium; the second condition means that the change in the optical path length caused by the passage of light through the particle is small. This implies that the field of the wave acting on each element of volume dV of the scattering particle does not differ appreciably, either in amplitude or phase, from the field of the incident wave. If $dV \ll (\lambda_0/m_0)^3$, then the amplitude functions for each element of volume can be found utilizing the theory of Rayleigh scattering, according to which

$$\left.\begin{array}{c} S_1(\theta) \\ S_2(\theta) \end{array}\right\} = ik^3\, \frac{m-1}{2\pi}\, e^{i\delta} dV \left\{\begin{array}{c} 1, \\ \cos\theta, \end{array}\right. \tag{2.7}$$

where the phase factor $e^{i\delta}$ allows for the phase shift between waves scattered by different volume elements of the particle arising from the different positions of these volume elements in space.

Allowance must be made for the interference of waves scattered by different volume elements. Accordingly, it is not the intensities of the waves scattered by different volume elements of the particle that must be added, but their amplitudes. In the case of scattering by a spherical particle, integration over volume gives

$$\left.\begin{array}{c} S_1(\theta) \\ S_2(\theta) \end{array}\right\} = ik^3 R^3 (m-1) \sqrt{\frac{2\pi}{u^3}}\, J_{1/2}(u) \left\{\begin{array}{c} 1, \\ \cos\theta. \end{array}\right. \tag{2.8}$$

Here $u = 2kR \sin(\theta/2)$ and $J_{3/2}(u)$ is a Bessel function. On inserting (2.8) into (2.6), we obtain, for the intensity of the scattered light,

$$I_p = \frac{4k^4 R^6}{9}\, |m-1|^2 G^2(u)\, [\cos^2\varphi + \cos^2\theta \sin^2\varphi]\, \frac{I_0}{r^2}, \tag{2.9}$$

where we have written $G(u) = (9\pi/2u^3)^{1/2} J_{3/2}(u)$.

It can be seen from (2.9) that the intensity of the scattered light is directly proportional to R^6 while the scattering indicatrix is determined by the function $G^2(u)$ and by the factor in the square brackets, which depends on the specifics of the experimental scheme used to measure the angular dependence of the intensity of the scattered light. Thus, if the scattering indicatrix is found experimentally, the radius of the scattering particle can be determined, as it enters as an unknown adjustable parameter into the function $G^2(u)$.

3. Scattering of Light by Electron — Hole Drops

In order to find the intensity of the light scattered by an electron—hole drop, we require to calculate the complex refractive index of the electron—hole liquid. If we are concerned with the scattering of electromagnetic waves for which the energy of a quantum is much greater than the binding energy of the particles in the liquid phase, then the carriers in the drops may be regarded as free. The real part of the dielectric constant of the electron—hole liquid then has the form [124]:

$$m_{Re}^2 = \varkappa_0 - \frac{4\pi e^2 n_0}{m^* \omega^2} \tag{2.10}$$

for $\omega \gg \gamma$. Here $\varkappa_0$ is the dielectric constant of the crystal, n_0 is the carrier density in the liquid phase, m^* is the reduced electron (hole) effective mass, ω is the angular frequency of the light wave, and γ is the collision frequency. Assuming that $4\pi e^2 n_0/(\varkappa_0 m^* \omega^2) \ll 1$, we obtain from (2.10)

$$m_{\text{Re}} = m_0 \left(1 - \frac{2\pi e^2 n_0}{\varkappa_0 m^* \omega^2}\right), \qquad (2.11)$$

since $m_0 = \sqrt{\varkappa_0}$. We can write, for the imaginary part of the refractive index of the drop,

$$m_{\text{Im}} = n_0 S \lambda_0/4\pi, \qquad (2.12)$$

where S is the cross section for absorption of light by free carriers. Then, utilizing (2.11) and (2.12), we obtain

$$|m - 1|^2 = \frac{4e^4 \lambda_0^2 + S^2 c^4 m^{*2} m_0^2}{16\pi^2 c^4 m^{*2} m_0^4} n_0^2 \lambda_0^2, \qquad (2.13)$$

where c is the free-space velocity of light.

Inserting (2.13) into (2.9) gives, for the intensity of the scattered light,

$$I_{\text{s}} = \frac{4\pi^2 n_0^2 R^6}{9c^4 m^{*2} \lambda_0^2} \left[4e^4 \lambda_0^2 + c^4 m^{*2} S^2 m_0^2\right] G^2(u) B \frac{I_0}{r^2}. \qquad (2.14)$$

Here we have introduced the notation

$$B = \cos^2 \varphi + \cos^2 \theta \sin^2 \varphi. \qquad (2.15)$$

In Eq. (2.14) the first term in the square brackets allows for the contribution from the real part of the refractive index to the scattering of light by the drop; the second term in the square brackets allows for the contribution from the imaginary part. At a wavelength $\lambda_0 = 3.39 \ \mu m$, the first term is greater than the second by a factor of around 8 for EHD in germanium. The limits of applicability of the Rayleigh–Gans theory can be estimated using (2.13); for EHD in germanium, this theory is valid for $R \ll 90 \ \mu m$.

Expression (2.14) gives the intensity of the light scattered by a single particle. The intensity of the light scattered at an angle θ by all electron–hole drops coming within the confines of the incident light flux is obtained by multiplying (2.14) by the total number N_D of drops which scatter light. If the experimental conditions are such that the cross section of the light beam S_0 is less than the dimensions of the region occupied by the drops, then $N_D = NS_0 d$, where N is the drop number density and d is the scattering length.

Concurrently with scattering, we may also measure the attenuation of the light flux transmitted through a sample containing nonequilibrium carriers bound into droplets and excitons. The attenuation comes both from absorption and from scattering of light by the particles. In our case, however, it can be shown that scattering makes only a small contribution to the attenuation. If we assume that the cross sections for absorption of light by particles bound into drops and excitons are the same, then we have for the power absorbed in the crystal:

$$W = SN_0 d\Phi_0, \qquad (2.16)$$

where Φ_0 is the incident light flux, and

$$N_0 = {}^4/_3 \pi R^3 n_0 N + n \qquad (2.17)$$

is the average density of nonequilibrium particles in the sample (i.e., sum of density of particles bound into drops and exciton density n).

At low temperatures, when $n_0 \gg n$,

$$W = {}^4/_3 \pi R^3 n_0 N S d \Phi_0. \tag{2.18}$$

Since $I_s \sim n_0^2 N$, where $W \sim n_0 N$, we see that scattering and absorption measurements provide a means of determining the particle density of the drops:

$$n_0 = \frac{3 c^4 m^{*2} \lambda_0^2 S}{\pi R^3 \left[4 e^4 \lambda_0^2 + c^4 m^{*2} S^2 m_0^2 \right] G^2 (u) B} \frac{I}{W}, \tag{2.19}$$

where $I = I_s N S_0 d$ is the intensity of the light scattered at an angle θ by all drops coming within the confines of the light flux, and we have used $\Phi_0 = I_0 S_0$. Having found n_0, the drop number density can be calculated using (2.18) or (2.14).

With increasing temperature, the excitons may begin to make a noticeable contribution to the absorption. In this case the absorption is described by Eqs. (2.17) and (2.16), and absorption and scattering measurements yield n_0 and N as a function of n. If, concurrently with absorption and scattering, we also observe EHD luminescence (from which the product $n_0 N$ can be determined, i.e., the total number of particles in the drops), then it becomes possible to determine all three quantities n_0, N, and n. However, since $I_s \sim R^6$, the error in the measurement of n_0 by this method is quite large, i.e., scattering measurements provide only an approximate estimate of the carrier density in the liquid phase.

4. Experimental Determination of EHD Parameters
by the Light Scattering Method

Scattering and absorption measurements were made using the setup shown in block diagram form in Fig. 3. This arrangement is similar in many ways to that described by Pokrov-

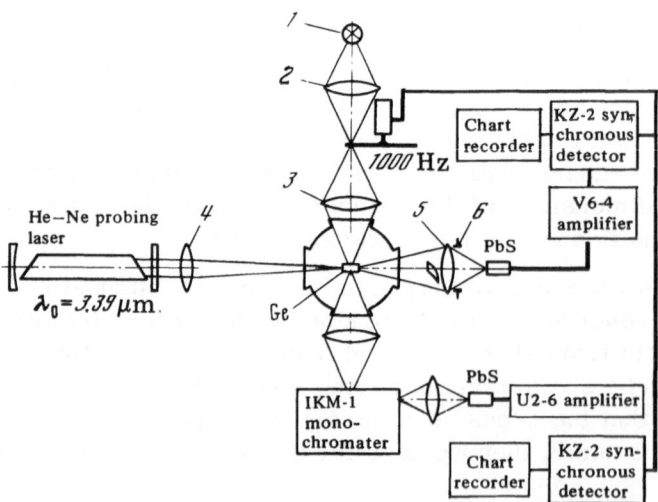

Fig. 3. Block diagram of experimental setup. (1) Incandescent lamp; (2, 3) high-power objectives for focusing exciting radiation onto sample; (4) lens for focusing probing laser radiation; (5) quartz lens for gathering scattered light; (6) diaphragm.

skii and Svistunova [52]. The source of excitation is an incandescent lamp 1 of nominal power 100 W. The filament of the lamp was chosen so as to reduce the relative contribution of the short-wavelength component of the exciting radiation. Light from the lamp was focused using two high-power objectives 2 and 3 onto a side face of the sample. The exciting radiation was modulated at 1 kHz using a mechanical chopper. The technique is thus a differential one, i.e., only the contribution of nonequilibrium carriers to the absorption or scattering of the probing radiation is recorded.

The probing radiation was provided by an LG-126 laser operating at a wavelength $\lambda_0 = 3.39~\mu m$. The parasitic 1.15 μm wavelength radiation which is always present in the output from this laser was eliminated by passing the output through a germanium light filter; the probing beam was then focused by a long-focus lens 4 upon the front face of the crystal so that, within the sample, it passed parallel to the side face illuminated by the incandescent lamp. The radiation emerging from the back face of the sample was focused by a quartz lens 5 onto a PbS receiver cooled to ~100°K. Stray illumination from the incandescent lamp and recombination luminescence from the crystal were prevented from entering the receiver by means of an interference light filter placed in front of the receiver. In the scattering measurements the direct laser beam was blocked by a screen placed in front of lens 5, the aperture angle of which was $2\theta_0 \simeq 10°$. The total aperture angle of the system for recording the scattering was $2\theta_1 = 22°$.

The angular distribution of the intensity of the scattered light was obtained by varying the aperture of diaphragm 6. The experimental results were analyzed using Eq. (2.14) with the factor B evaluated in accordance with our experimental arrangement. The light flux scattered within the incremental solid angle bounded by cones with apex angles 2θ and $2(\theta + d\theta)$ (here and below θ is the scattering angle in vacuo; in the above formulas θ denoted the scattering angle in the crystal) is given by [125]

$$d\Phi(\theta) = \frac{8\pi^3 R^6 n_0^2 Nd~[4e^4\lambda_0^2 + m^{*2}c^4 m_0^2 S^2]}{9\lambda_0^2 m_0^2 m^{*2} c^4}~\Phi_0 G^2(u)\sin\theta d\theta. \qquad (2.20)$$

In the derivation of this expression the right side of (2.14) was integrated with respect to φ from 0 to 2π and allowance was made for the change in the solid angle due to refraction at the crystal boundary and the inclined incidence of the scattered light on lens 5. It is also assumed that $\cos\theta_{cr} \simeq 1$, as the scattering angle in the crystal θ_{cr} is small. The experimental results were analyzed on the assumption that there is no change in the scattering volume [126], since under the conditions of our experiment the relevant corrections are negligible. In this manner, by measuring the angular distribution of the intensity of the scattered light, we obtained the experimental dependence on θ of the function $\Phi(\theta) = \int_{\theta_0}^{\theta} d\Phi(\theta)$. This function, suitably corrected for the dependence of the transmission of the interference light filter on angle of incidence, is shown in Fig. 4.

The procedure used to measure the Ge recombination luminescence was standard, and employed an IKM-1 monochromator and a cooled PbS receiver. Electrical signals were recorded using a conventional low-frequency apparatus with synchronous detection.

The samples, of dimensions $15 \times 5 \times 2$ mm, were soldered into the bottom of the helium vessel of the cryostat so that one-half of the sample was in the helium and the other half, the working part, was in vacuo. The measurements were made on mechanically polished Ge samples with a residual impurity content of less than 10^{13} cm^{-3}.

Scattering measurements were made using the sample with the greatest exciton lifetime. In this sample, for excitation by light from the incandescent lamp, the exciton density measured

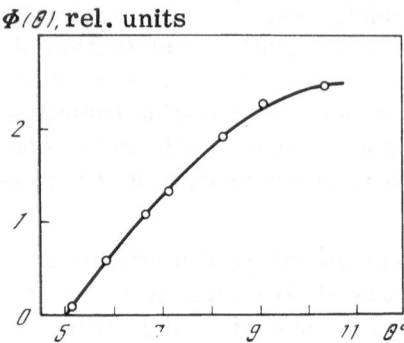

$\Phi(\theta)$, rel. units

Fig. 4. Dependence of Φ on angle θ measured
at T = 2.8°K.

via the absorption of 3.39 μm probing radiation decreases slowly with increasing distance from
the illuminated surface, the density at a distance of 1 mm from the illuminated surface being
less than the density at 0.5 mm by a factor of only 1.7, i.e., excitation is effectively a volume
effect (Fig. 5). The integrated scattering signal (i.e., the scattering signal for diaphragm 6
completely open) was the greatest for this sample.

5.   Results and Discussion

It can be seen from (2.20) that the quantity $(1/\sin\theta)[d\Phi(\theta)/d\theta]$ is the function $G^2(u)$ in
arbitrary units. Thus, by graphically differentiating the plot of $\Phi(\theta)$ shown in Fig. 4 and dividing
the result by $\sin\theta$, we obtain the experimental values of the function $G^2(u)$. The parameter R
was adjusted to give the best fit between the experimental points and the curve of G(u) calculated theoretically (Fig. 6). The EHD radius found in this manner was R = 10 μm at T = 2.8°K,
in reasonable agreement with the results of [52].

Having determined the radius of the drops, it is possible to find the EHD carrier density.
To this end we utilize (2.19), suitably modified to incorporate (2.20). We find after some algebra that

$$n_0 = \frac{3\lambda_0^2 m_0^2 m^{*2} c^4 S}{2\pi^2 R^3 [4e^4\lambda_0^2 + m^{*2}c^4 m_0^2 S^2]\, W G^2(u)}\; \frac{1}{\sin\theta}\, \frac{d\Phi(\theta)}{d\theta}. \tag{2.21}$$

If we set S = 2 · 10^{-16} cm^2 (which corresponds to the cross section for absorption by holes in
Ge in interband transitions [127]) and m* = $m_e m_h/(m_e + m_h)$ = 0.083 m_f where m_f is the free-
electron mass, m_e = 0.12 m_f is the effective electron mass, m_h = 0.27 m_f is the effective electron mass entering into the expression for the hole part of the electrical susceptibility [128],
then the carrier density in an electron−hole liquid found from the experimental data using (2.21)

Fig. 5. Distribution of absorption over sample
thickness h measured from illuminated surface
for excitation by incandescent lamp. (1) T = 4.2° K;
(2) T = 2.1°K. Threshold temperature $T_0 \simeq 3.4°$ K.

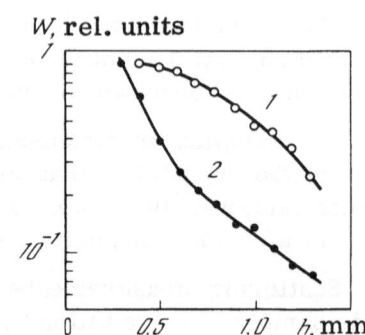

W, rel. units

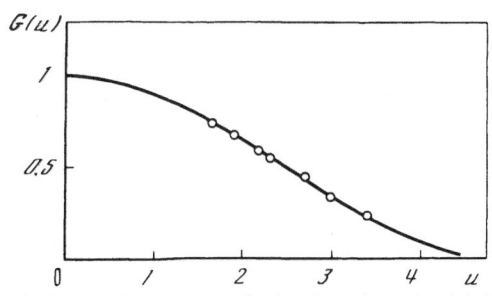

Fig. 6. Dependence of G on u. The solid curve is the theoretical function G(u). The experimental points correspond to R = 10 μm at T = 2.8°K.

is $n_0 \simeq 1.5 \cdot 10^{17}$ cm^{-3} [53]. This value is in good agreement with the results of other papers [6, 9, 19, 37, 67, 91].

It should be noted that, according to [129], some doubt attaches to the use of Eq. (2.10) for germanium, since it does not allow for interband hole transitions. However, in our experiments $\hbar\omega$ is greater than the spin—orbit splitting of the valence band, and the formulas of [129], derived ignoring the third band, will not be applicable. A quantitative treatment of the experimental data allowing for interband transitions was therefore not undertaken. It may be that allowing for interband transitions will slightly reduce the value of n_0 determined from our experiments. Also, it is known that the absorption cross section of photoexcited carriers may be different from the absorption cross section of free carriers measured using doped samples [130]. The value of S cited in [55] is $3 \cdot 10^{-17}$ cm^2. If this value is employed in the treatment of our results, however, the value obtained for n_0 will be too small.

Having determined the radius of the drops and the carrier density in them, the number of drops per unit volume of the crystal N can be found using (2.18) or (2.20). Under our experimental conditions, the EHD number density is $N \simeq 10^5$ cm^{-3}.

A study of the dependence on temperature of the absorption, the integrated scattering, and the recombination luminescence [53] showed that at the lowest working temperatures, when the photoluminescence spectrum contains only the EHD line, the absorption is almost independent of temperature whereas the integrated scattering signal increases with increasing temperature (Fig. 7). As temperature is further increased, the exciton line appears in the recombination spectrum and the absorption signal starts to decrease. The integrated scattering signal passes through a maximum, and falls to zero at a temperature ~3.4°K. At this temperature only the exciton line remained in the photoluminescence spectrum, while the absorption signal stopped falling and thereafter hardly varied with increasing temperature up to 4.2°K.

The growth of the integrated scattering signal is connected with an increase in the drop radius with temperature (the cross section for integrated scattering is proportional to R^4 [123]). As temperature is further increased up to the condensation threshold, the drop number density

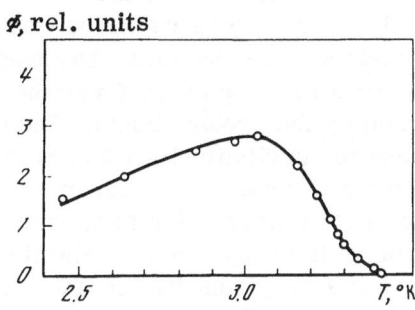

Fig. 7. Dependence of integrated scattering signal Φ on temperature.

falls sharply and along with it the volume of the liquid phase. The integrated scattering signal consequently decreases.

CHAPTER III

EFFECT OF TEMPERATURE ON EXCITON CONDENSATION CONDITIONS IN GERMANIUM

The strong temperature dependence of the integrated scattering signal described in the previous chapter indicates that drop size is essentially dependent on temperature. It is of interest to investigate in more detail the dependence of drop radius on temperature, as the dimensions of the drops can be expected to depend on their number density and the latter must depend on how drops nucleate.

The setup on which our first measurements were made was, however, not suitable for detailed systematic observations: due to insufficient sensitivity, the angular distribution of the intensity of the scattered light could be measured only near the maximum of the integrated scattering curve; inaccuracy in positioning the screen limiting the aperture of the system for small-angle scattering introduced an error into the measurement of the angular distribution; at large excitation levels the light from the incandescent lamp overheated the sample; the measurements and the processing of the results were very time-consuming.

In view of these shortcomings, we constructed a high-sensitivity setup incorporating automatic recording of the angular distribution of the intensity of the scattered light. High sensitivity was attained through using a laser amplifier to amplify the scattered light. The source of exciting light was a laser operating at 1.52 μm wavelength, which created a sufficiently uniform distribution of nonequilibrium carriers over the volume of the sample.

In the first part of this chapter we describe the experimental setup in some detail and the preparation of the samples; in the second, we cite and discuss the experimental results.

1. High-Sensitivity Setup for Studying

Light Scattering by EHD in Ge

A block diagram of the experimental setup is shown in Fig. 8 [131]. The source of exciting radiation is a helium−neon laser operating in the cw mode at 1.52 μm wavelength. This laser was constructed around an LG-36A industrial laser. An output power of ~15 mW was attained by suitably choosing the discharge tube and the mirrors. Maximum power was attained with a multimode tube and an output mirror with a reflection coefficient of 95%.

The coefficient of absorption of 1.52 μm light in Ge is not large, amounting to ~9 cm^{-1} at liquid helium temperature. Such a source of excitation thus gives a distribution of nonequilibrium carriers bound into drops and excitons that is quite uniform over the thickness of the sample. This distribution was measured using the scheme shown in Fig. 9. The absorption of 3.39 μm probing radiation was measured as a function of the distance h from the illuminated surface of the sample. The magnitude of the absorption of the 3.39 μm radiation is proportional to the average (over the sample volume) carrier density in that part of the sample traversed by the probing beam. The distribution of the absorption of 3.39 μm light over sample thickness for excitation by 1.53 μm laser radiation is shown in Fig. 10. It can be seen that this distribution depends on temperature and excitation intensity (for an exciting power $P \simeq 8$ mW, the threshold condensation temperature $T_0 \simeq 4.2°$K). However, even when the absorption vs. h distribution is pressed toward the illuminated surface of the sample, the carrier density at a distance of 1 mm from this surface is only a factor of 1.5 smaller than near it.

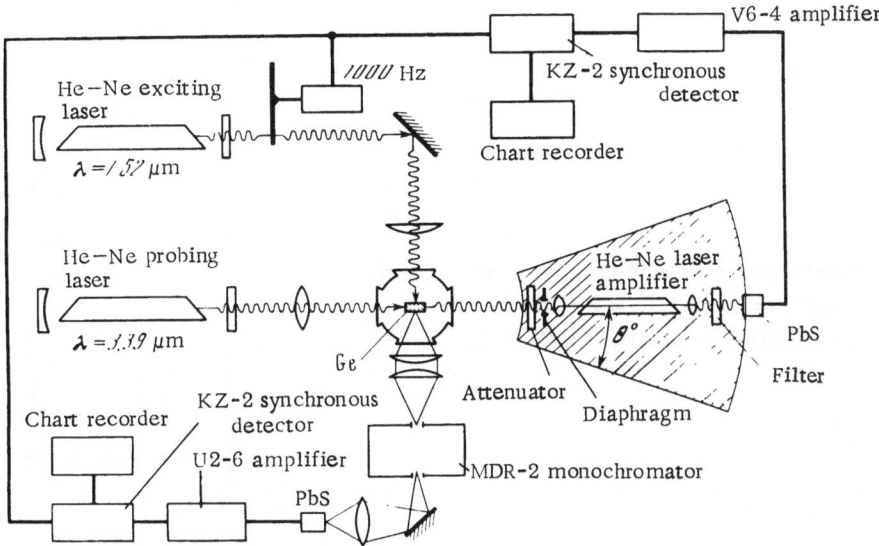

Fig. 8. Block diagram of setup for investigating light scattering
by EHD in germanium.

The exciting radiation was focused onto the side face of the sample by a cylindrical lens into a narrow strip parallel to the 3.39 μm probing laser beam.

In the experiments we measured the absorption and scattering of 3.39 μm wavelength probing radiation from a helium−neon laser. An LG-36A laser was reconstructed for operation at this wavelength. A narrow single-mode beam of power ~20 mW was obtained by using a single-mode tube and an internal diaphragm. Greatest output power was obtained by using, as the output mirror, a polished Ge plate or the wide-band mirror from an LG-126 laser. It should be noted that the intensity of the scattered light cannot be increased to any significant extent by increasing the power of the probing laser, since, if the intensity of the scattered light is sufficiently great, a reduction of the EHD luminescence intensity is observed [133].

The probing beam was focused onto the front surface of the sample by a long-focus lens into a spot of diameter ~300 μm in such a manner that the beams of the two lasers were superposed. Superposition was judged in terms of maximum absorption signal. The beam being scattered passed at a distance of 0.6-1.5 mm from the illuminated surface of the sample. The scattered light emerged through the back face of the crystal, and its angular distribution was automatically recorded by means of a goniometer on which were mounted a laser amplifier and a photoreceiver. The moveable part of the goniometer was displaced by means of a synchronous motor in the horizontal plane at a rate of 2.16 deg/min along an arc of a circle centered on the investigated sample. The center of rotation of the goniometer was situated near the back face of the sample; the position of the center of rotation in a direction perpendicular to the 3.39 μm probing beam was checked via the symmetry of the scattering indicatrix.

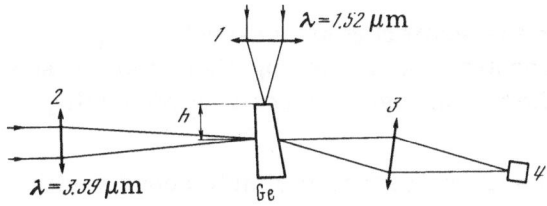

Fig. 9. Measurement of absorption vs. distance h from illuminated surface. (1) Lens focusing exciting radiation; (2) lens focusing probing radiation; (3) lens focusing probing radiation passed through sample onto photoreceiver (4).

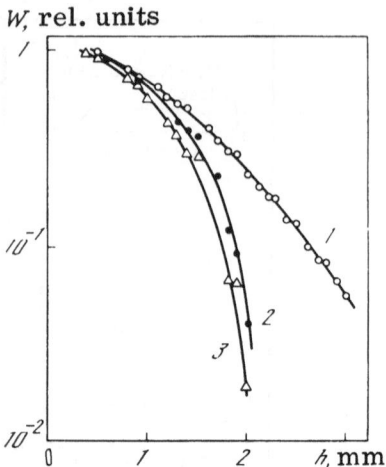

Fig. 10. Distribution of absorption over sample thickness. (1) P = 8 mW, T = 2.8°K; (2) P = 2.4 mW, T = 2.8°K; (3) P = 8 mW, T = 3.8°K.

An electrical discharge in a gaseous helium−neon mixture has a gain coefficient for 3.39 μm radiation that is quite large [134], an effect which was utilized to enhance the sensitivity of the scattering measurements [133]. The laser amplifier had the form of the discharge tube from an LG–55 laser. The relevant parameters are: tube diameter ~2 mm; length of discharge gap ~190 mm; discharge current ~15 mA. The gain coefficient at 3.39 μm was greatest at this current, and for various samples of discharge tubes lay in the range W_{out}/W_{in} = 12–18 for input powers W_{in} < 10 μW.

An optical attenuator and a stack of diaphragms were placed in front of the laser amplifier. The diaphragms defined the light flux scattered at a definite angle and determined the angular resolution of the system. The attenuator reduced the light flux by a factor of ~10^4. It was used when the absorption signal or the intensity distribution in the probing laser beam was being recorded, since the gain coefficient of the amplifier is reduced for large input intensities, amounting to ~3 at an input power of ~1 mW. The radiation passed through the diaphragms was collimated by a quartz lens, amplified by the amplifier, and then focused onto a liquid-nitrogen-cooled PbS photoreceiver. A narrow-band light filter transmitting in the 3 μm region was placed in front of the PbS receiver to prevent radiation from the amplifier discharge from falling upon the receiver. The amplifier had an adjustment mechanism whereby its tube could be aligned along the laser beam shone through the sample.

In the measurements of the angular distribution of the light scattered by the drops, the angular resolution of the system was varied from 0.25 to 0.5° depending on the magnitude of the signal. The angular resolution could be improved for large signals by reducing the diameter of the diaphragms positioned before the laser amplifier. To illustrate the operation of the setup, we show in Fig. 11 part of the recording of the diffraction pattern obtained on diffraction of the 3.39 μm light by a 1-mm-diameter circular aperture substituted for the sample. The recording was obtained at an angular resolution of ~5'; the laser beam was focused on the amplifier input diaphragm. Maxima from the 7th to the 14th are shown in the figure; however, a total of around 100 maxima could be recorded.

The exciting radiation was modulated at 1 kHz in the scattering and absorption experiments, i.e., the technique is a differential one; the absorption and scattering signals are respectively in antiphase and in-phase with the excitation. Electrical signals were recorded using a standard low-frequency setup.

The spectrum of the Ge recombination luminescence was recorded simultaneously with the scattering and absorption. The photoluminescence was observed using a conventional setup

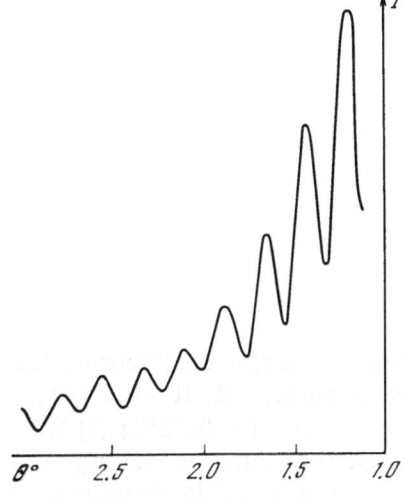

Fig. 11. Part of recording of diffraction pattern
from a circular aperture.

in which the mononchromator was a high-transmission MDR-2. The radiation was detected by
a PbS receiver cooled to ~100°K.

In order to determine the radius of the drops from the angular distribution of the intensity
of the scattered light, we require a suitable expression describing this distribution in which
due allowance is made for our experimental setup. In our measurements the angular distribution was recorded in the horizontal plane, while the laser tube was oriented so that this plane
is also the plane of polarization for the linearly polarized laser radiation; accordingly, in
Eq. (2.15) the angle $\varphi = 90°$ and B = 1. Utilizing (2.14) and allowing for refraction at the crystal
boundary, we then obtain for the light flux scattered into the aperture of our recording system
at an angle θ to the probing laser beam:

$$I_s = \frac{4\pi^2 n_0^2 N d R^6}{9 c^4 m^{*2} \lambda_0^2 m_0^2} [4 e^4 \lambda_0^2 + c^4 m^{*2} S^2 m_0^2] G^2(u) \Phi_0 \Omega, \qquad (3.1)$$

where Ω is the solid angle subtended at the center of rotation of the goniometer by the amplifier
aperture diaphragm. Thus, when the angular distribution is recorded using our goniometer setup, we see from this expression that the quantity recorded in arbitrary units on the penrecorder
chart is $G^2(u)$ as a function of the scattering angle.

A recording of the absorption signal for one of the germanium samples is shown in Fig.
Fig. 12a. The shape of this signal reproduces the intensity distribution in the 3.39 μm laser
beam. A recording of the angular distribution of the intensity of the light scattered by EHD is

Fig. 12. Recordings of absorption and scattering signals. (a) Recording of absorption signal;
(b) angular distribution of intensity of light
scattered by EHD. R = 9.5 μm; T = 3.23°K.

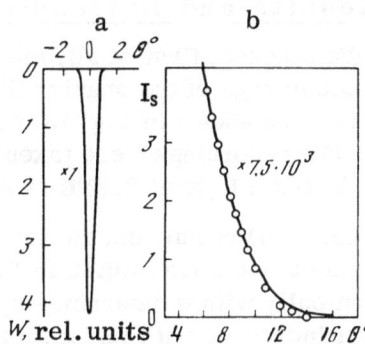

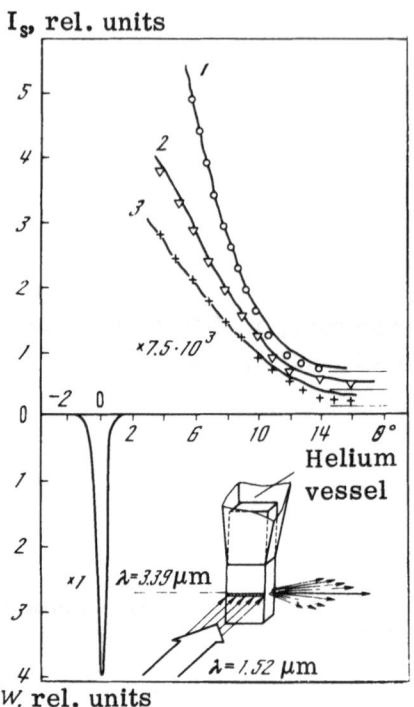

Fig. 13. Recordings of angular distribution for three different temperatures; (1) R = 9.5 mW, T = 3.23°K; (2) R = 8.5 mW, T = 3.12°K; (3) R = 8 mW, T = 3.06°K. Bottom left: recording of absorption signal taken at T = 3.12°K; bottom right: passage of radiation through sample.

shown in Fig. 12b. The points correspond to calculated values of the function $G^2(u)$ for drops of radius R = 9.5 μm.

The measurements were made on mechanically polished Ge samples of residual-impurity concentration ~10^{12} cm^{-3} and ~10^{11} cm^{-3}. The face of the sample through which the scattered light emerged was inclined at an angle of 2° to the opposite face to eliminate parasitic interference. The samples, of dimensions 15 × 5 × 2 mm, were soldered into a copper cone, which was in turn soldered with Wood's alloy to the helium vessel of the cryostat and served as the bottom of the helium vessel (Fig. 13). The copper cone was made by electrolytic deposition of copper onto a Duralumin former. It was made sufficiently thin (wall thickness was not more than 300 μm) to reduce the mechanical stresses arising in the sample during cooling due to the different thermal expansions of Ge and Cu, since it is known [7] that electron−hole drops can move away from the region of excitation to the region of greatest strain. That half of the sample that was to be located in the liquid helium was first ground and then electrolytically provided with a thin layer of copper. The copper was then tinned with indium, and the sample soldered with indium to the cone so that the solder seam passed only along the line of contact of the cone and the sample.

2. Results and Discussion

We discuss, first of all, the results obtained from the samples with $N_d - N_a$ ~ 10^{12} cm^{-3} [54]. Recordings of the angular distribution of the scattered light taken at three different temperatures are shown in Fig. 13; the points correspond to the calculated value of the function $G^2(u)$. The recordings were taken at an angular resolution of 0.5°. A recording of the absorption of 3.39 μm light at 3.12°K is shown at the bottom left of the figure.

The EHD radius and the intensity I_s of light scattered by EHD at angle of 8° (in vacuo) to the incident beam are plotted in Fig. 14 as functions of temperature. The EHD radius increases monotonically with temperature up to the threshold temperature (~3.5°K). The scattering signal first increases with temperature, passes through a maximum, and then decreases.

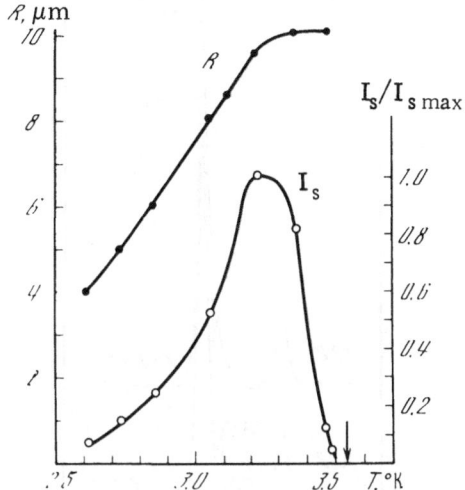

Fig. 14. EHD radius R and intensity of scattered light I_s as functions of temperature. The arrow indicates the threshold temperature.

The intensities (at line peaks) of the FE and EHD luminescence lines are plotted in Fig. 15 as functions of temperature along with the absorption and scattering signals. It can be seen that with increasing temperature from 2 to ~3°K the absorption signal remains almost unchanged, while the intensity of the EHD line falls by a factor of around 20%. The more rapid (compared with the absorption) decrease of the EHD line in this range of temperatures is probably connected with the fact that the distribution over sample thickness of nonequilibrium carriers bound into holes is temperature-dependent (see Fig. 10). With further increase of temperature the absorption and EHD signals begin to fall sharply, while the FE line increases. At a temperature of ~3.53°K the scattering signal vanishes and only the FE line remains in the photoluminescence spectrum. Spectra of the Ge recombination luminescence taken at increasing temperatures are shown in Fig. 16.

The mean density of nonequilibrium carriers in that part of the sample traversed by the 3.39 μm probing beam was estimated from the absorption of this radiation and equaled ~ $3 \cdot 10^{14}$ cm^{-3} at 2.6°K; the value was less than this by a factor of around 30, however, at 3.6°K, when all the nonequilibrium carriers are bound into excitons. This reduction of the mean nonequilibrium-carrier density with increasing temperature is explained by the smaller lifetime of

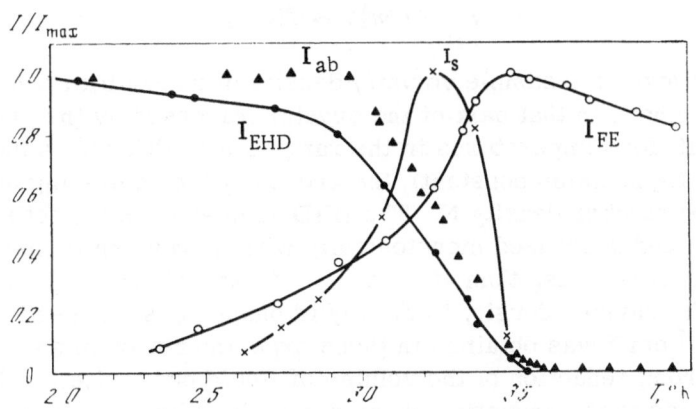

Fig. 15. Temperature dependence of absorption I_{ab}, scattering I_s, and intensities of FE and EHD lines I_{FE} and I_{EHD} respectively.

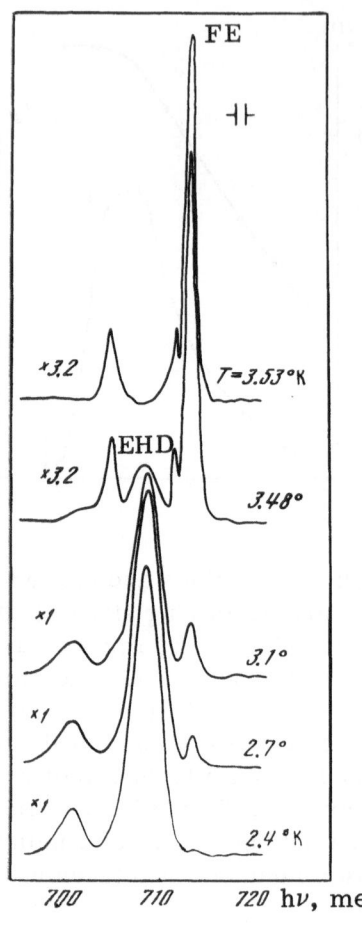

Fig. 16. Spectra of Ge recombination luminescence taken at various temperatures.

carriers bound into excitons as opposed to carriers bound into holes, and also by the increase
in the volume of the region in which nonequilibrium carriers are localized that accompanies
the transition through the threshold temperature as a result of exciton diffusion [77, 125].

By (3.1), the intensity of the light scattered by the electron–hole drops is:

$$I_s \sim R^6 n_0^2 N \sim N_\Sigma^2/N, \tag{3.2}$$

where N_Σ is the mean (over the sample volume) density of nonequilibrium carriers bound into
droplets of condensed phase in that part of the crystal traversed by the probing beam. It can
be seen from (3.2) that, for temperatures in the range 2.6–3.0°K, where the absorption signal
does not change (i.e., N_Σ remains constant), the growth of the scattering signal must be due to
a reduction of the drop number density N. The EHD number density, estimated via (3.1), was
$\sim 6 \cdot 10^6$ cm^{-3} at 2.6°K and decreased monotonically with increasing temperature; near the
condensation threshold it was less than at 2.6°K by a factor of around 100. In these estimates
it was assumed that the carrier density in the liquid phase n_0 is independent of temperature;
its value $n_0 = 1.9 \cdot 10^{17}$ cm^{-3} was obtained in these experiments from absorption and scattering
data at 2.6°K. The strong reduction of the volume of liquid phase (i.e., of N_Σ) near the condensation temperature leads to the reduction of the scattering signal as the threshold is approached.

In this manner, the number of "working" condensation centers decreases with increasing
temperature, i.e., evidently, condensation occurs on "deeper" centers. Thus, for an unchanging
or even diminishing mean (over sample volume) density of carriers bound into drops, which is

determined by the form of the exciton gas—electron-hole liquid phase diagram constructed allowing for the nonequilibrium character of this system, it follows that the size of the drops must increase with increasing temperature.

If we assume that excitons condense not at condensation centers but as a result of fluctuations of exciton density (this situation can obtain in an ideal crystal, where there are no condensation centers), the increase of drop size with temperature can be understood on the basis of known ideas on the formation of nuclei of "critical" size [16, 135].

We note that at large scattering angles the calculated points do not coincide with the experimental scattering curves (see Fig. 13), i.e., the latter possess "tails." These tails are evidently connected with the presence in the sample of drops of dimensions different from the mean, the difference being the greater the lower the temperature. With increasing temperature the exchange of particles between the liquid and gaseous phases intensifies and equalization of the sizes of the drops becomes possible, which leads to the better agreement between the calculated points and the experimental curves at higher temperatures.

The radius of the drops was measured as a function of the distance between the probing laser beam and the illuminated surface of the sample. These measurements showed that as this distance is increased from 0.6 to 1.5 mm, the drop radius decreases by not more than 10-15%, the change being greater the lower the temperature. A stronger dependence of radius on distance was observed in [52]; however, when an incandescent lamp is used as the source of excitation [52], carriers bound into drops will be distributed in a more nonuniform manner over the thickness of the sample than for excitation by 1.52 μm laser light. With increasing temperature, when the absorption vs. distance distribution is compressed toward the illuminated surface (see Fig. 10), the drop radius, as will be shown in the following chapter, is weakly dependent on excitation intensity. Accordingly, the reduction of the volume of liquid phase with increasing distance from the illuminated surface at temperatures close to threshold is explained by a reduction of drop number density.

In conclusion, we stop briefly on the results obtained from the samples with an impurity content of ~10^{11} cm^{-3}. The experimental conditions were the same as described above. In accordance with the results of [22], the threshold temperature of condensation for these samples was around 0.4°K lower than for the samples with impurity content ~10^{12} cm^{-3} for the same excitation level. The drop radius varied more weakly with temperature: for a change of temperature from 2.7 to 3.1°K (threshold temperature $T_0 \simeq 3 \cdot 15$°K), the radius of the drops increased from 10 to 12 μm. The mean (over sample volume) density of carriers bound into electron—hole drops amounted to $N_\Sigma \simeq 1.2 \cdot 10^{14}$ cm^{-3} at $T = 2.7$°K and the drop number density was ~$1.5 \cdot 10^5$ cm^{-3}. As the threshold temperature is approached, the drop number density fell to ~$3 \cdot 10^3$ cm^{-3}. It would appear that the number of condensation centers in these samples is small, so that the EHD number density is small as well. The small (compared with the less pure samples) drop number density leads to an increase in the drop radius; and, since the steady-state density of the exciton gas increases as the radius of the drops increases, it follows that, for a given generation rate, the volume of liquid phase must be reduced.

CHAPTER IV

THE KINETICS OF EXCITON CONDENSATION

The results described in Chapter III show that the EHD number density, which is determined by the conditions and mechanism of nucleation, is strongly dependent on the experimental conditions. This implies that the volume of condensed phase may differ greatly from the vol-

ume for thermodynamic equilibrium and is primarily determined by the number of drops that
are formed [56]. Accordingly, considerable interest attaches to a detailed study of the dependence of the EHD size and number density on experimental conditions: temperature, excitation
intensity, rise time of the exciting light pulse. In the present chapter we cite the results of
such experiments. The results are discussed on the basis of a theory of condensation in which
allowance is made for the diffusion of excitons towards the surface of the drops and for the
surface tension of the electron−hole liquid. Allowance for diffusion and surface tension greatly
changes the rate equations describing the growth and recombination of liquid-phase drops [3,
19] used in a whole string of papers to analyze experimental results [19, 24, 46, 81, 86, 110].
These refinements of the theory proved, however, to be insufficient to explain the behavior of
the drop radius at high temperatures. Accordingly, we consider the effect on the kinetics of
droplet growth of exciton−exciton collisions and the removal of excitons by the phonons produced as a result of recombinations in the drops.

On the assumption that exciton condensation occurs as a result of density fluctuations
in the exciton gas, it can be shown that the coefficient of surface tension of the electron−hole
liquid can be found from the temperature dependence of the EHD number density; the coefficient of surface tension has been calculated in a number of papers [11-14]. The results of
measurements of the coefficient of surface tension are cited in the present chapter, and we
discuss the fluctuational mechanism of formation of nuclei of the liquid phase. It should be
noted that this mechanism is apparent only in sufficiently pure crystals. The condensation
centers in doped semiconductors are probably bound multiexciton complexes [20, 21, 24, 33,
105-107]; even here, however, exciton density fluctuations are necessary in order that a complex grow into a nucleus of "critical" size.

1. Kinetics of Condensation and Steady-State EHD Size

If any sort of matter finds itself in a metastable state, then it must go over into another
state − the stable state. For a system of excitons, such a metastable state is a state in which
the exciton "vapor" is supersaturated. Let us consider the transition of a supersaturated
exciton vapor to the stable state, i.e., to a state in which part of the vapor is condensed into
a liquid and the ratio of the volumes of the liquid and gaseous phases is determined by an
appropriate phase diagram. The transition process begins with the formation of nuclei of the
liquid phase. Nuclei can be formed only as a result of fluctuations of the vapor density or
because molecules of vapor are able to accumulate near condensation centers (for example,
excitons near defects of the crystal structure). In order that a spherical nucleus of radius R
be in equilibrium with the gas, the density of the gas above the surface of the nucleus must be
(ignoring recombination in the liquid phase) [135]:

$$n_\tau (R) = n_{0\tau} \exp (2\sigma/n_0 kTR), \qquad\qquad (4.1)$$

where σ is the coefficient of surface tension of the electron−hole liquid, n_0 is its density, and
[3]

$$n_{0\tau} = \nu \left(\frac{M^* kT}{2\pi\hbar^2}\right)^{3/2} \exp\left(-\frac{\varepsilon_0}{kT}\right) \qquad\qquad (4.2)$$

is the thermodynamic-equilibrium vapor density above a "flat" liquid surface. In (4.2) the
quantity $\nu = 16$ is the degeneracy factor; $M^* = 0.335\,m_f$ is the effective density-of-exciton-
states mass; m_f is the free−electron mass; $\varepsilon_0 = 2.1$ meV [37] is the binding energy per pair
of particles in the liquid reckoned from the ground exciton level (numerical values here and
everywhere below are cited for Ge).

It can be seen from (4.1) that, for a given vapor density, only nuclei of a definite radius R* can be in equilibrium with the vapor. If the radius of a nucleus is less than R*, the vapor pressure proves to be too small for equilibrium to be possible and such a nucleus rapidly evaporates. If, however, the radius is greater than R*, the nucleus begins to grow. In this manner, nuclei of radius R* find themselves in a state of unstable equilibrium with the vapor. Such nuclei are called critical nuclei.

The fluctuational probability of formation of a nucleus of "critical" size is given by [135]

$$W \sim \exp\left(-A/kT\right). \tag{4.3}$$

where A is the minimum work required to form a nucleus of radius R*. The quantity A is given by

$$A = 4\pi\sigma R^{*2}/3. \tag{4.4}$$

Inserting (4.4) into (4.3), we obtain, for the fluctuational probability of formation of a nucleus of "critical" size,

$$W \sim \exp\left(-4\pi\sigma R^{*2}/3kT\right). \tag{4.5}$$

The dependence of the "critical" radius on temperature and on the density n of the supersaturated vapor is given by (4.1) rewritten in the form

$$R^* = \frac{2\sigma}{n_0 kT \ln\left(n/n_{0T}\right)} . \tag{4.6}$$

Inserting (4.6) into (4.5) gives, for the fluctuational probability of formation of "critical" nuclei,

$$W \sim \exp\left[-\frac{16\pi\sigma^3}{3n_0^2\,(kT)^3 \ln^2\left(n/n_{0T}\right)}\right]. \tag{4.7}$$

This expression is convenient for analyzing the dependence of the rate of formation of nuclei on supersaturated vapor density n at constant temperature. If the vapor density is kept constant, we can use (4.2) to determine the threshold temperature T_0 for the given vapor density and a flat interface between phases with the aid of the relationship:

$$n = v\left(\frac{M^* kT_0}{2\pi\hbar^2}\right)^{3/2} \exp\left(-\frac{\varepsilon_0}{kT_0}\right). $$

Replacing n in (4.6) by this expression and n_{0T} by (4.2), we obtain an equation describing the temperature dependence of the radius of a "critical" nucleus:

$$R^* = \frac{2\sigma}{n_0\varepsilon_0\left[(T_0 - T)/T_0 + 3/2\,(kT/\varepsilon_0) \ln\left(T_0/T\right)\right]} . \tag{4.8}$$

Expression (4.5) then gives

$$W \sim \exp\left\{-\frac{16\pi\sigma^3}{3n_0^2 e_0^2 kT\,[(T_0 - T)/T_0 + 3/2\,(kT/\varepsilon_0) \ln\left(T_0/T\right)]^2}\right\}. \tag{4.9}$$

Expression (4.9) gives the probability of formation of "critical" nuclei as a function of temperature T for a given supersaturated vapor density.

If a nucleus of liquid phase is formed of size slightly greater than "critical," then this nucleus will start to grow. So long as the dimensions of the nucleus are small compared with the exciton mean free path, it grows because the excitons, in their random thermal motion, impinge by chance upon the surface of the nucleus and are trapped by it. However, when the dimensions of the growing drop become comparable with the exciton mean free path, excitons which find themselves near the surface of the drop are rapidly captured by this surface, with the result that the exciton density around the drop falls. A diffusion flux of excitons from the depths of the crystal towards the surface of the drops thus arises.

Let us consider the growth of an electron–hole drop until it reaches a state of equilibrium with the surrounding exciton gas [58]. The time variation of the number of particles in a spherical drop of radius R is described by

$$\frac{d}{dt}\left(^4/_3\pi R^3 n_0\right) = S - {}^4/_3\pi R^3 \frac{n_0}{\tau_0}, \tag{4.10}$$

where τ_0 is the carrier lifetime in the liquid phase, and

$$S = 4\pi R^2 \left[n(R) - n_\text{т}(R)\right] v_\text{т} \tag{4.11}$$

is the flux of excitons onto the surface of the drop. In (4.11) the quantity n(R) is the exciton density near the surface of the drop, and $v_\text{T} = (kT/2\pi m^*_\text{ex})^{1/2}$ is the thermal velocity, which determines the number of exciton impacts with the surface of the drop (m^*_ex is the effective exciton mass). The term $4\pi R^2 n_\text{T}(R) v_\text{T}$ describes evaporation of excitons from the surface of the drop. Expression (4.11) differs from that used for the exciton flux in [3, 19, 24, 46, 81, 86, 110]; in all of these papers n(R) was taken to be the mean exciton density in the sample volume, and n_OT given by (4.2) was used in place of the quantity $n_\text{т}(R)$, i.e., diffusion of excitons toward the drop surface and the surface tension of the electron–hole liquid were ignored. Condensation kinetics were analyzed in [103] using a statistical approach based on calculation of the probabilities of capture and emission of excitons by liquid-phase drops. The effect of surface tension on the energy of evaporation of excitons from the drop was taken into account in this paper, but the difference between n(R) and n was ignored. It will be shown below, however, that for sufficiently large R the rate of growth of the drops is limited by the rate of diffusion of excitons towards the drop surface, while the creation of a diffusion flux requires an exciton density gradient in the vicinity of the drop.

The dependence of the exciton density n(r, t) on the coordinate r and distance t is determined by the continuity equation

$$\frac{\partial n(\mathbf{r}, t)}{\partial t} + \text{div}\left[-D\nabla n(\mathbf{r}, t)\right] + \frac{n(\mathbf{r}, t)}{\tau} = g(\mathbf{r}, t), \tag{4.12}$$

where D is the exciton diffusion coefficient, τ is the exciton lifetime, and g(r, t) is the rate of generation of nonequilibrium carriers by the source of excitation (it is assumed that the time required for carriers to bind themselves into excitons is much less than all the other characteristic times of the problem). Expression (4.11) at the surface of each drop constitutes the boundary condition on (4.12). Taking the distance between drops to be much less than the exciton diffusion length $L_\text{D} = (D\tau)^{1/2}$, we can average Eq. (4.12) over volume elements which contain many drops but which are small compared with L_D. This gives

$$\frac{\partial \bar{n}(\mathbf{r}, t)}{\partial t} + \text{div}\left[-D\nabla \bar{n}(\mathbf{r}, t)\right] + NS(\bar{n}, R) + \frac{\bar{n}(\mathbf{r}, t)}{\tau} = g(\mathbf{r}, t), \tag{4.13}$$

where $\bar{n}(\mathbf{r}, t)$ denotes the exciton density averaged in this manner, and N is the drop number density. The derivation of (4.13) utilizes

$$\overline{\mathrm{div}\,[-\,D\nabla n\,(\mathbf{r},\,t)]} = \mathrm{div}\,[-\,D\nabla\bar{n}\,(\mathbf{r},\,t)] + NS\,(\bar{n},\,R),$$

i.e., this sort of averaging enables us to regard as separate the diffusion of excitons toward the surfaces of the drops [the term $NS(\bar{n}, R)$] and into the bulk of the crystal (the term $\mathrm{div}[-D\nabla\bar{n}(\mathbf{r}, t)]$). If the excitation of the sample is spatially uniform, $\mathrm{div}\,[-D\nabla\bar{n}(\mathbf{r}, t)] = 0$ and the diffusion fluxes of excitons are directed solely toward the drops. Equation (4.13) describes the variation in time of the mean exciton density $\bar{n}(\mathbf{r}, t)$ allowing for their generation, recombination, capture by and evaporation from EHD, and diffusion.

In order to find the flux $S(\bar{n}, R)$, which depends on the mean exciton density $\bar{n}$ and the EHD radius R, we are required to solve (4.12) in the vicinity of each given drop subject to boundary conditions (4.11) and $n(r \to \infty) = \bar{n} \equiv n$ (the size of the drops is assumed small compared with the distance between them.[†] Since we are now talking about solving the problem in a small region $r \ll N^{-1/3}$ (r is distance measured from the center of the drop) within which a quasi-stationary exciton density is established in a very short time, the terms $\partial n/\partial t$, n/τ, and g can be omitted in (4.12), when it goes over into the steady-state diffusion equation

$$\frac{D}{r^2}\frac{d}{dr}\left[r^2\,\frac{dn\,(r)}{dr}\right] = 0. \tag{4.14}$$

Since, by definition of the diffusion coefficient,

$$S = 4\pi R^2 D\,\frac{dn\,(r)}{dr}\Big|_{r=R},$$

the solution of Eq. (4.14) allowing for the boundary conditions has the form

$$n\,(r) = n - S/4\pi Dr. \tag{4.15}$$

Combining this solution with (4.11), we obtain

$$S\,(n,\,R) = 4\pi R^2\,\frac{n - n_{\mathrm{T}}\,(R)}{1 + v_{\mathrm{T}}R/D}\,v_{\mathrm{T}}. \tag{4.16}$$

Inserting S(n, R) from (4.16) into (4.10) and (4.13), we obtain equations describing the growth and recombination of electron–hole drops and the kinetics of the gas phase for a known EHD number density:

$$\frac{dR\,(\mathbf{r},\,t)}{dt} = \frac{\bar{n}\,(\mathbf{r},\,t) - n_{\mathrm{T}}\,(R)}{n_0}\,\frac{v_{\mathrm{T}}}{1 + v_{\mathrm{T}}R\,(\mathbf{r},\,t)/D} - \frac{R\,(\mathbf{r},\,t)}{3\tau_0}, \tag{4.17}$$

$$\frac{\partial\bar{n}\,(\mathbf{r},\,t)}{\partial_t} + \mathrm{div}\,[-\,D\nabla\bar{n}\,(\mathbf{r},\,t)] + 4\pi R^2\,(\mathbf{r},\,t)\,N\,(\mathbf{r})\,\frac{\bar{n}\,(\mathbf{r},\,t) - n_{\mathrm{T}}\,(R)}{1 + v_{\mathrm{T}}R\,(\mathbf{r},\,t)/D}\,v_{\mathrm{T}} + \frac{\bar{n}\,(\mathbf{r},\,t)}{\tau} = g\,(\mathbf{r},\,t). \tag{4.18}$$

For spatially uniform excitation ($\nabla n = 0$) and small R and $\sigma = 0$, Eqs. (4.17) and (4.18) go over

---

[†] Below, where there is no chance of confusion, the bar above n will be omitted; we shall, of course, understand by n the exciton density averaged in the above manner.

into the rate equations that are normally used [3, 19, 24, 46, 81, 86, 110]:

$$\frac{dR}{dt} = \frac{n - n_{\text{or}}}{n_0} v_{\text{T}} - \frac{R}{3\tau_0} \tag{4.19}$$

and

$$\partial n/\partial t + 4\pi R^2 N (n - n_{0\text{T}}) v_{\text{T}} + n/\tau = g(t). \tag{4.20}$$

However, when the dimensions of the drops become comparable with the exciton mean free path $(v_{\text{T}}R/D \gtrsim 1)$, the strong reduction of exciton density near the drops compared with the mean volume exciton density n means that subsequent drop growth is greatly slowed down. In the case of very small drops, Eq. (4.17) gives a drop growth rate that is much less than (4.19), since, by (4.1), the presence of surface energy increases the rate of evaporation of excitons from the drop surface.

Under steady-state conditions, the flux of excitons onto the surface of the drop equals the rate at which carriers recombine within it:

$$S = {}^4/_3\pi R^3 n_0/\tau_0. \tag{4.21}$$

Inserting (4.21) into (4.15), we obtain, for the density distribution of the exciton gas in the vicinity of an electron—hole drop,

$$n(r) = n - n_0 R^3/3D\tau_0 r, \tag{4.22}$$

where n is given by (4.17) with dR/dt = 0, i.e.,

$$n = n_{\text{T}}(R) + \frac{n_0 R}{3v_{\text{T}}\tau_0}\left(1 + \frac{v_{\text{T}}R}{D}\right). \tag{4.23}$$

The function n(r) is plotted in Fig. 17 for R = 4 μm and T = 2°K. The calculations were performed using (4.22), (4.23), (4.1), and (4.2) and the following numerical values of the relevant quantities: $n_0 = 2 \cdot 10^{17}$ cm^{-3}, $\tau_0 = 20$ μsec, D = 1500 cm^2/sec at 3°K [77] (for scattering on acoustic phonons we took D ~ T$^{-1/2}$), $m_{\text{ex}} = 0.4\,m_f$, and $\sigma = 1.8 \cdot 10^{-4}$ dyn/cm. It can be seen from (4.22) and Fig. 17 that the exciton density changes considerably over a distance of the order of a few R from the surface of the drop. Far from the drop, the density of the exciton gas equals the mean exciton density in the volume of the crystal and is given by expression (4.23).

Expression (4.23) describes the dependence of the steady-state radius of an electron—hole drop on the mean density of the exciton gas. Plots of n vs. R calculated using (4.23) for

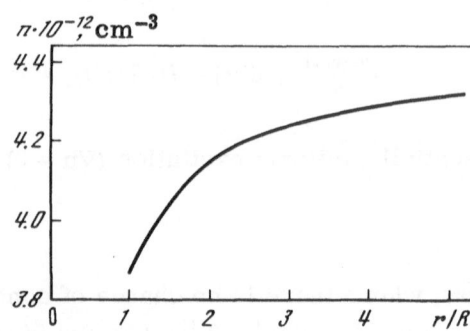

Fig. 17. Distribution of exciton density n(r) near drops of liquid phase. T = 2°K; R = 4 μm.

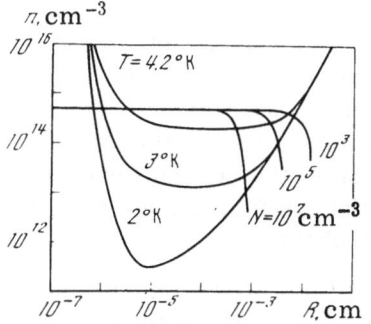

Fig. 18. Plots of n vs. R calculated using (4.23) for three different temperatures and using (4.26) for three different EHD number densities N. $\tau = 5$ μsec and g = 10^{20} cm^{-3}sec^{-1}; numerical values of other parameters are cited in the text.

three different temperatures are shown in Fig. 18. Allowing for surface tension has the effect that, as the drop radius decreases, the exciton density n does not tend towards n_{OT} but instead grows exponentially. The steady-state solution for small R leads to expression (4.6) for the radius of the "critical" nucleus. An interesting feature of these curves is the appearance of a minimum of exciton density. The position of the minimum is determined by the intersection of the two branches of the function n: the unstable branch, describing nuclei of "critical" size; and the stable branch (R > R_{min}), determining the equilibrium size of the drops. Thus, at a given temperature, only drops of radius not less than the minimum can be in stable equilibrium with the exciton gas.

The radius of drops of minimum size is readily found for the case where R $\ll$ D/v_T and $2\sigma/n_0RkT \ll 1$. Differentiating (4.23) with respect to R and equating the result to zero, we obtain

$$R_{min} = \frac{1}{n_0} \sqrt{\frac{6\sigma v_T n_{OT}}{kT}} . \tag{4.24}$$

Inserting (4.24) into (4.23) gives the least steady-state exciton density at which electron–hole drops are capable of existing in stable equilibrium with the exciton gas:

$$n_{min} = n_{OT} + 2\sqrt{2\sigma n_{OT}/3v_T\tau_0kT} . \tag{4.25}$$

For a steady spatially uniform excitation ($\partial n/\partial t = 0$, $\nabla n = 0$, g(r, t) = g = const), simultaneous solution of (4.18) and (4.23) gives the condition for the conservation of the number of particles in the considered nonequilibrium system:

$$g = \frac{n}{\tau} + \frac{4\pi R^3 n_0 N}{3\tau_0} , \tag{4.26}$$

i.e., the rate of generation equals the sum of the rates of recombination in the liquid and gaseous phases. If this expression is used to plot n as a function of R on the graph of Fig. 18, then the point of intersection of the curves obtained using (4.23) and (4.26) will give for small R the "critical" radius of the nuclei for the given generation rate and, for large R, the steady-state value of the drop radius for the given generation rate and a particular drop number density.

The simultaneous solution of Eqs. (4.23) and (4.26) can be written in the form

$$g - \frac{n_{OT}}{\tau} = \frac{n_0R}{3v_T\tau_0\tau}\left(1 + \frac{v_TR}{D}\right) + \frac{n_{OT}}{\tau}\left(e^{2\sigma/n_0RkT} - 1\right) + \frac{4\pi R^3 n_0 N}{3\tau_0} . \tag{4.27}$$

This expression gives the amount by which the rate of generation has to exceed the thermodynamic equilibrium threshold for a "flat" gas—liquid interface, n_{OT}/τ, in order to maintain stable equilibrium in a system consisting of an exciton gas and electron—hole drops of radius R or to form a "critical" nucleus of given size. The first term on the right side of (4.27) gives the amount by which the exciton density has to exceed the thermodynamic equilibrium density so that the additional flux of excitons to the drop surface compensates the reduction in the volume of the drop due to carrier recombination in the liquid phase; the second term allows for the surface energy of the drops and determines the behavior of the function $g - n_{OT}/\tau$ at small R; the last term gives the rate of recombination in the liquid phase.

The function $g - n_{OT}/\tau$ is plotted in Fig. 19 for T = 2°K and N = 10^7 and 10^5 cm^{-3}, and for T = 4.2°K and N = 10^4 and 10^3 cm^{-3}. The calculations were carried out for the same numerical values of the parameters as for the curves of Figs. 17 and 18. The two horizontal straight lines correspond to generation rates of 10^{20} and $3 \cdot 10^{19}$ cm$^{-3} \cdot$ sec^{-1}. The points where these straight lines intersect the graphs of the function $g - n_{OT}/\tau$ give, for small R, the "critical" radius for the given generation rate, and, for large R, the steady-state drop radius for the given generation rate and the particular drop number density.

We note that the positions of the minima of the curves given by (4.23) and (4.27) coincide only when the rate at which carriers recombine in drops of minimum radius is negligible compared with the rate at which they recombine in the gaseous phase. Thus, for example, for $R \ll D/v_T$ and $2\sigma/n_0 kTR \ll 1$, the positions of the minima coincide when

$$N \ll \frac{m_{ex}^* n_0^2}{144\tau_0 \tau s n_{OT}}. \qquad (4.28)$$

In the opposite case of a large drop number density, the minima of the curves shown in Fig. 19 are shifted in the direction of small R and their position depends on the drop number density. In this case

$$R_{min}^* = \sqrt[4]{\frac{\sigma\tau_0 n_{OT}}{2\pi N n_0^2 \tau kT}}, \qquad (4.29)$$

and the value of the function $g - n_{OT}/\tau$ at its minimum does not determine the smallest generation rate at which EHD can be formed; at this generation rate all that is possible is the formation of "critical" nuclei of radius R_{min}^*, which are in unstable equilibrium with the exciton gas.

In the case when the drop number density is small, i.e., when inequality (4.28) is fulfilled, it is an easy matter to find the least generation rate required to form drops of minimum size

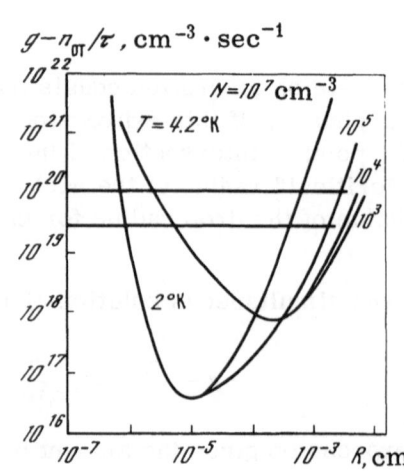

Fig. 19. Plots of $g - n_{OT}/\tau$ vs. R calculated using (4.27).

in stable equilibrium with the exciton gas. To this end we require to insert R_{min} from (4.24) into (4.27) and neglect the last term or divide (4.25) by τ, remembering that in this case $(g - n_{OT}/\tau)_{min} = (n_{min} - n_{OT})/\tau$. We find for the relative minimum excess above threshold:

$$\frac{(g - n_{OT}/\tau)_{min}}{n_{OT}/\tau} = 2\sqrt{\frac{2\sigma}{3 v_T \tau_0 k T n_{OT}}}, \qquad (4.30)$$

i.e., the relative excess above threshold increases with decreasing temperature, since n_{OT} then decreases exponentially. The increase in the relative minimum supersaturation of the exciton gas with decreasing temperature is connected with the decrease of the minimum stable drop radius, as a result of which the contribution of the surface energy to the energy of evaporation of the drops is increased. This effect, the increase in the relative minimum supersaturation with decreasing temperature, explains the results of [31, 46], where the temperature dependence of the threshold excitation intensity was found to deviate at low temperatures from the dependence predicted by (4.2), and also the results of [30], where the optical hysteresis observed by the authors is more sharply expressed at low temperatures than at high.

We conclude this section by writing out the equations by means of which the steady-state EHD radius may be determined. The experimentally observed EHD radii are much greater than R_{min}; accordingly, in the calculation of the steady-state radius of the drops, we may neglect the contribution of the surface energy to the evaporation energy, i.e., we may neglect the second term on the right side of (4.27). The drop radius is then given by the solution of a cubic equation. Let us consider a number of particular cases with a view to obtaining simpler formulas for R. If the recombination rate in the liquid phase is much greater than the exciton recombination rate (low temperatures and a high excitation level), the last term on the right side of (4.27) is much greater than the rest, and

$$R = \sqrt[3]{\frac{3\tau_0}{4\pi n_0} \frac{g}{N}}. \qquad (4.31)$$

i.e., the EHD radius is proportional to $(g/N)^{1/3}$.

For the opposite relationship between the recombination rates:

$$R = -\frac{D}{2 v_T} + \sqrt{\frac{D^2}{4 v_T^2} + \frac{3\tau_0 D}{n_0}(g\tau - n_{OT})}. \qquad (4.32)$$

In the case of sufficiently small drops $(R \ll D/v_T)$, Eq. (4.32) gives

$$R = \frac{3 v_T \tau_0}{n_0}(g\tau - n_{OT}). \qquad (4.33)$$

This formula describes the dependence of drop radius on temperature and excitation intensity in those cases when the generation level is not much above threshold.

In the opposite case (large drops):

$$R = \sqrt{\frac{3 D \tau_0}{n_0}(g\tau - n_{OT})}. \qquad (4.34)$$

It can be seen from Eqs. (4.32)–(4.34) that if recombination goes primarily through the gaseous phase, the radius of the drops does not depend on their number density. We note that as the condensation threshold is approached, when the radius of the drops decreases and tends toward R_{min} corresponding to the threshold temperature, the contribution of the surface energy

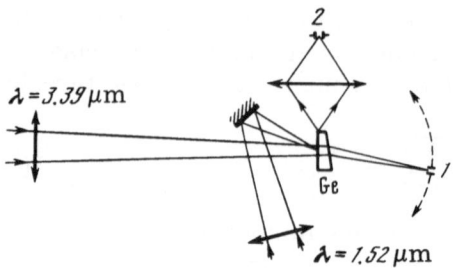

Fig. 20. Scheme used for measurement of total number of scattering particles. (1) Inlet diaphragm of laser amplifier; (2) monochromator slit.

may become significant, in which case n_{OT} in expressions (4.32)–(4.34) must be replaced by $n_T(R)$.

2. Experimental

The experimental technique is essentially the same as described previously (Chapter III, Section 1). However, in addition to the scheme shown in Fig. 13, the laser beams were also superposed as shown in Figs. 20 and 21. These schemes were utilized when it was necessary to increase the generation rate. The EHD radius and number density were measured as functions of excitation intensity with the exciting light sharply focused on the front face of the crystal and probing beam focused on the input diaphragm of the laser amplifier (Fig. 20). The beams of the two lasers were superposed in terms of a maximum of absorption signal; superposition was checked via the symmetry of the diffraction pattern from the region occupied by carriers bound into EHD (Fig. 22). The dimensions of this region can be estimated from the angular position of the diffraction minima [125]; its diameter is ~600 μm for sharp focusing of the exciting radiation. This is around three times greater than the diameter of the spot into which the exciting light was focused, measured using a phosphor sensitive to 1.52 μm light [136].

With the probing radiation focused onto the input diaphragm of the laser amplifier, the diameter of the 3.39 μm laser beam on the sample was greater than the diameter of the region occupied by drops of condensed phase; accordingly, it was possible to determine from the scattering measurements not the number density of the drops but the total number of drops in the sample. Due to the variation with temperature and excitation intensity of the dimensions of the region in which the EHD are located, the total number of drops in the sample does not need to vary in proportion to their number density. The drop number density was thus determined in this case from the EHD luminescence signal, the intensity of which is proportional to the total number of particles in the liquid phase in the volume defined by the image of the monochromator slit on the sample. Thus, having determined the radius of the drops from the angular distribution of the scattered light, the total number of drops in this volume in relative units could be found. The width of the slit image on the sample was a few times less than the thickness of the sample. Accordingly, the averaging in the luminescence measurements is

Fig. 21. Scheme used for measurement of number density of scattering particles. (1) Inlet diaphragm of laser amplifier; (2) monochromator entrance slit.

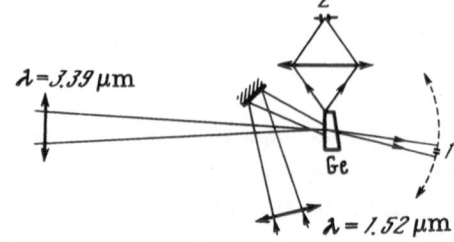

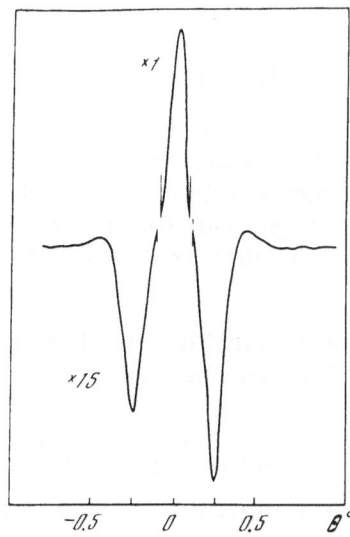

Fig. 22. Diffraction pattern of 3.39 μm light
from region occupied by EHD.

over a smaller volume of the sample than in the scattering measurements, with the result that variations in the dimensions of the region occupied by the drops are less apparent.

The dependence of the EHD radius and number density on excitation intensity was measured by inserting calibrated light filters into the 1.52 μm laser beam. Light pulses with a steeper front were produced by sharply focusing the exciting radiation onto the disk of the mechanical chopper.

In order to determine the coefficient of surface tension of the electron–hole liquid, it was necessary to measure the temperature dependence of the drop number density. In these experiments the laser beams were superposed as shown in Fig. 21. The exciting radiation was focused onto the front face of the sample into a spot of diameter 0.35 mm. The probing beam was also focused onto this surface; the diameter of the probing beam was ~300 μm (at the focus). The closeness of the diameters of the two beams meant that they had to be superposed with particular care; otherwise diffraction from the edge of the region occupied by the drops was observed, which made quantitative measurements difficult. As the diameter of the probing beam is less than the diameter of the region of excitation in measurements with this superposition scheme, scattering measurements yield the drop number density averaged over the thickness of the sample.

All experiments described in this chapter were carried out on samples of pure germanium with a residual-impurity density of less than 10^{12} cm^{-3}. The samples were prepared as described in Section 1 of Chapter III.

3. Surface Tension of the Electron – Hole Liquid

Expressions (4.7) and (4.9) show that the probability of formation of nuclei of liquid phase is essentially dependent on the coefficient of surface tension of the electron–hole liquid. The coefficient of surface tension can be estimated using Langmuir's formula [16],

$$\sigma \simeq \varepsilon_0 n_0^{2/3}/5, \tag{4.35}$$

which is valid for liquids. This formula has a simple physical significance: the latent heat of evaporation referred to unit volume of liquid equals the combined surface energy of all the molecules, as if this volume of liquid were divided into individual molecules. An estimate of the coefficient of surface tension using (4.35) gives $\sigma \sim 2 \cdot 10^{-4}$ dyn/cm. Rigorous theoretical calculations of this quantity have been carried out in a number of papers to date [11–14], according to which $\sigma = (0.8\text{–}1.5) \cdot 10^{-4}$ dyn/cm. Until now, the coefficient of surface tension has not been determined experimentally.

The light-scattering method provides an opportunity of estimating σ experimentally. According to (4.9), the probability of formation of "critical" nuclei W is a function of temperature and is strongly dependent on σ. If we assume that the drop number density is proportional to W, then, by measuring the temperature dependence of the drop number density and applying (4.9), we can determine the coefficient of surface tension.

The measurements were performed using the scheme shown in Fig. 21. The drop number density was found from scattering, absorption, and EHD luminescence data. Excitation was by pulses of front duration less than 3 μsec.

The dependence of the EHD number density N on the quantity $\frac{1}{\theta} = \left[T\left(\frac{T_0 - T}{T_0} + {}^{3}/_{2}\frac{kT}{\varepsilon_0}\ln\frac{T_0}{T}\right)^2\right]^{-1}$. is shown in Fig. 23. It can be seen that at low temperatures the experimental points lie rather well along a straight line. From (4.9), the slope of this line can be used to determine the coefficient of surface tension. If we take $\varepsilon_0 = 2.1$ meV and $n_0 = 2 \cdot 10^{17}$ cm^{-3}, then we obtain $\sigma = 1.8 \cdot 10^{-4}$ dyn/cm [15, 58], in reasonable agreement with the theoretical calculations.

It should be noted that the threshold temperature determined experimentally is probably less than the true thermodynamic-equilibrium threshold temperature, since near the threshold the degree of supersaturation of the exciton gas is small and condensation may not occur, even though the exciton density exceeds the least density for which drops are capable of being formed. Also, at low temperatures, when the probability of formation of "critical" nuclei is large, the excitons produced when the excitation is switched on may start to condense before their density reaches the maximum value determined by the excitation intensity; the conditions of applicability of (4.9) will thereupon be violated. These two factors introduce an error into the determination of σ by this method. Excitation by short light pulses would eliminate the second factor but would increase the error in the determination of the threshold temperature.

4. Results and Discussion of Experiments on Condensation Kinetics

The results of the experiments described in Chapter III and a theoretical study of the kinetics of condensation show that at low temperatures and high excitation levels the EHD radius depends strongly on the drop number density. Accordingly, it would be interesting to

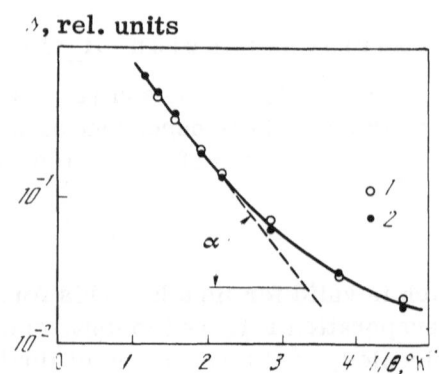

Fig. 23. Graph of EHD number density vs. $1/\theta$.
(1) From luminescence data; (2) from scattering data; $\tan \alpha = 16\pi\sigma^3/6.9 k n_0^2 \varepsilon_0^2$; threshold temperature $T_0 = 4.4°$K.

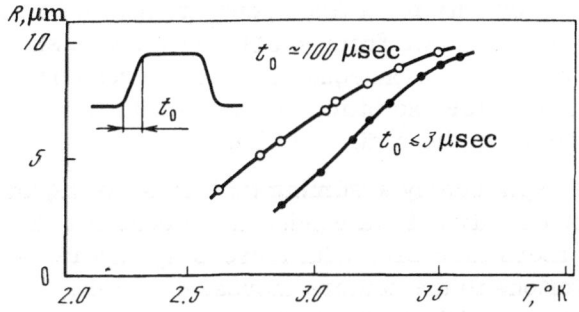

Fig. 24. Dependence of EHD radius on temperature measured using exciting pulses with long
and short fronts.

alter the nucleation conditions without changing the temperature or excitation intensity. This
can be done using pulses of quasistationary excitation, i.e., pulses of equal amplitude and
duration but different rise rates.

The EHD radius and number density were measured for excitation by pulses with long
($t_0 \simeq 100$ μsec) and short ($t_0 \lesssim 3$ μsec) fronts using the scheme shown in Fig. 13. The dependence of drop radius on temperature measured for these two cases is shown in Fig. 24. Steepen-
ing the front reduces the drop radius, the difference between the radius values determined for
long and short fronts increasing with decreasing temperature.

The drop number density was determined from scattering and absorption measurements.
Near the threshold temperature, the drop number densities for long and short fronts were
much the same; with decreasing temperature, however, the drop number density for the short
front increased more rapidly (Fig. 25).

These results are consistent with the picture of condensation kinetics considered in Section 1 of the present chapter. When the excitation intensity increases rapidly (t_0 comparable
with the exciton lifetime), the maximum exciton density possible for the given generation level
($n \sim g\tau$) is created in the sample, after which the process of condensation begins. In the case
where the exciting pulse has a long front, however, condensation begins when the exciton density reaches n_{min} (see Fig. 18). For a steep front, condensation begins at a greater exciton
density; thus, compared with the case of a long front, the "critical" radius [Eq. (4.6)] is less
and the probability of formation of "critical" nuclei [Eq. (4.7)] is greater, and therefor also

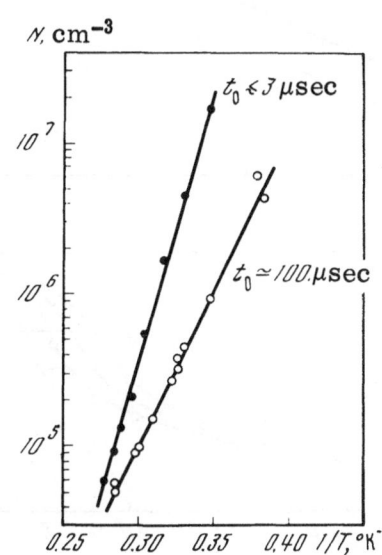

Fig. 25. Temperature dependence of drop number density measured using exciting pulses with
long and short fronts.

the drop number density. Increasing the drop number density for a fixed generation rate leads to a reduction of the steady-state radius of the drops, as follows from (4.31). With increasing temperature the quantity n_{min} increases and tends toward g_T. Consequently, with increasing temperature, the difference in the drop number densities for excitation by pulses with long and short fronts will be smaller, and thus the difference in the radii will also.

We note that the above results can readily be explained by assuming that nuclei of liquid phase are formed not through fluctuations of exciton density but via condensation centers. In this case, the number of "working" condensation centers increases with increasing supersaturation of the exciton "vapor," since the flux of excitons to the centers increases and the nuclei attain "critical" dimensions even for centers for which the cross section for exciton capture is small.

We consider now the experiments on the dependence of the EHD number density and steady-state radius on excitation intensity. It should be noted that these dependences may be different for excitation by pulses with long and short fronts due to the different nucleation conditions in these two cases.

These experiments were carried out using the scheme shown in Fig. 20. The drop number density was calculated from the intensity of the EHD luminescence line.

Plots of drop radius vs. temperature for two different excitation intensities measured using an exciting pulse with a long front are shown in Fig. 26a. It can be seen that reducing the excitation level reduces the drop radius, in agreement with the results of [52]. The opposite situation obtains, however, when the experiments are performed using an exciting pulse with a short front (Fig. 26b): here the radius of the drops increases when the excitation intensity is reduced.

Plots of drop number density vs. temperature measured for the same excitation intensities as the radius—temperature curves of Fig. 26 and for short and long pulse fronts are shown in Fig. 27 in relative units. The absolute values of the total number of electron—hole drops and the total number of electron—hole pairs in the sample, estimated from absorption and scattering measurements, are equal, for maximum excitation level, to $\sim 10^4$ and $\sim 10^{12}$ re-

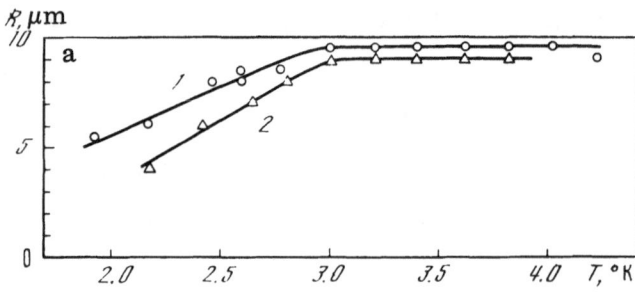

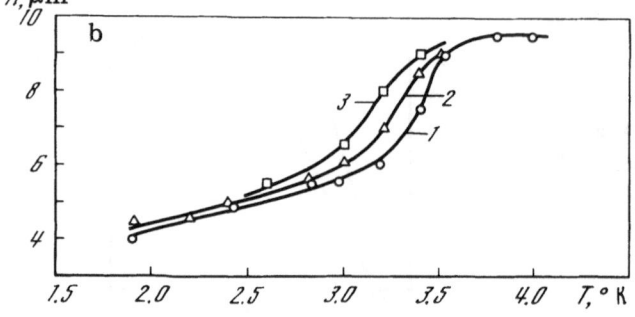

Fig. 26. Temperature dependence of EHD radius for various excitation intensities. (a) Long front; (b) short front. (1) $P \simeq 8$ mW; (2) 3.8 mW; (3) 2.4 mW.

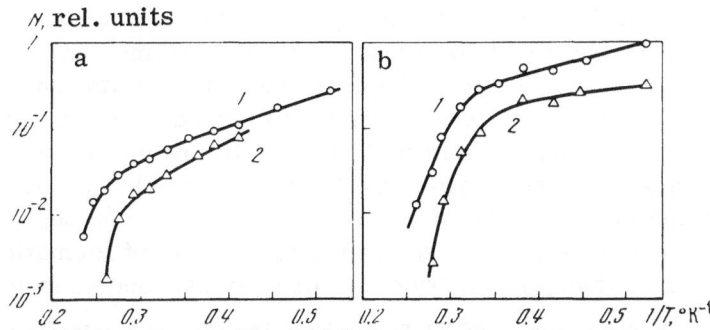

Fig. 27. Temperature dependence of EHD number density for two different excitation intensities. Excitation by pulses with (a) a long front; (b) a short front. (1) P ≃ 8 mW; (2) 3.8 mW.

spectively at 1.9°K; their number densities are equal to ~2 · 10⁷ and ~2 · 10¹⁵ cm⁻³ respectively. The number density of the drops and their total number are cited for excitation by pulses with a long front.

Plots of EHD number density vs. excitation intensity for various temperatures measured using long and short fronts are shown in Fig. 28. For a given temperature, the EHD number density decreases more slowly with decreasing generation rate for a long front (Fig. 28a) than for a short front (Fig. 28b). All curves become steeper with increasing temperature.

The stronger dependence of EHD number density N on excitation intensity g for excitation by pulses with a short front comes about because, in this case, the drops begin to grow when the exciton density reaches its maximum value, determined by the generation level; thus, reducing the excitation intensity leads to growth of the "critical" radius and so also to a reduction of W, the probability of formation of nuclei. With increasing temperature the curves of N vs. g become steeper since, for a given density of the exciton gas, the derivative $|dR^*/dn|$ increases with increasing temperature, as is apparent from Fig. 18.

In the case of exciting pulses with a long front, condensation begins when the amount by which the generation level exceeds threshold reaches a definite value (i.e., definite for each

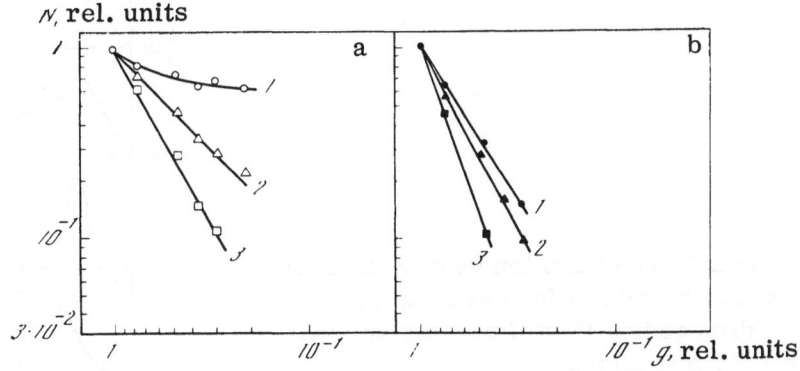

Fig. 28. Dependence of EHD number density N on excitation level g at three different temperatures. Excitation by pulses with (a) a long front; (b) a short front. (1) T = 2.6°K; (2) 3.2°K; (3) 3.6°K.

temperature) (see Fig. 19). Further increase of the excitation intensity should, generally speaking, result only in the growth of the nuclei formed at the moment condensation begins, and then the drop number density would be independent of the excitation intensity. Experimentally, however, the EHD number density is observed to increase slightly (compared with the short-front case), which probably comes about because the rate of formation of "critical" nuclei for small excesses of the generation level above threshold is small. With decreasing temperature, the minima of the curves of Figs. 18 and 19 are displaced in the direction of smaller R; this leads by (4.5) to an increase in the probability of formation of "critical" nuclei, and the dependence of drop number density on generation rate becomes weaker.

The results on the dependence of EHD number density on excitation intensity can be explained, along the same lines as previously, by assuming that the number of "working" condensation centers varies with excitation level.

In this manner, we have been observing phenomena of hysteresis type; i.e., despite the fact that the duration of the front, over which the excitation was varying in time, was not more than 20% of the time for which excitation was completely steady, the number density and dimensions of the electron—hole drops depended nonetheless on the slope of the front of the exciting pulse. This means that the EHD or their nuclei are created entirely at the beginning of the pulse; thereafter, during all the remainder of the pulse, the number of EHD scarcely changes. Qualitatively, this result can be accounted for in the following manner. The exciton density in the sample begins to increase from the moment of application of an exciting pulse with a front of increasing intensity (Fig. 29). When the exciton density noticeably exceeds $n_T \simeq n_{min}$, EHD nuclei begin to be formed; however, to start with, the number of excitons absorbed into these nuclei is relatively small, since both the number of nuclei and the radius of each of them are small, so that the term $NS \sim NR^2$ in Eq. (4.13) plays no role. The exciton density thus continues to grow (for a time less than τ, or, in the case of a long front, over a time less than the duration of the front). However, with increasing degree of supersaturation of the exciton gas, the number of newly formed nuclei increases rapidly (exponentially) and the radii of previously formed drops increase in accordance with (4.17). At some moment of time the number of excitons going into the drops becomes comparable with the rate at which carriers are being generated, and subsequently indeed exceeds it. From this moment the exciton density begins

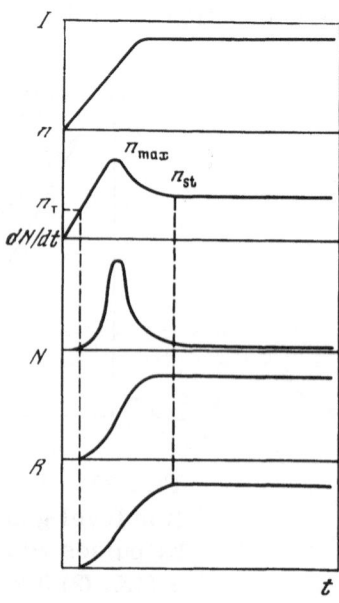

Fig. 29. Dependence of exciton density n, EHD number density N, rate of formation of nuclei dN/dt, and drop radius R on time t for an exciting pulse I of finite slope.

to fall, despite the constant or even still increasing excitation intensity I, since all the excitons being created by the excitation go toward the growth of previously nucleated electron—hole drops. The rate at which nuclei are being formed, dN/dt, thereupon also begins to decrease, and soon falls almost to zero when the exciton density drops to a level close to n_{st}, at which the probability of formation of nuclei is vanishingly small. The EHD previously formed still continue to grow for a certain time until they reach a size determined by Eqs. (4.23) and (4.26), the number of drops remaining fixed over this time.

This picture, based on the stabilization of the exciton density at a level n_{st}, close to n_{OT}, seems at first glance to be at variance with certain known results [30, 86, 108], which show that even under steady-state excitation conditions (especially at temperatures T > 3°K) the number of excitons continues to increase with increasing excitation level after the condensation threshold has been traversed. A more likely explanation of this fact, however, is that excitation is not spatially uniform in any of the experiments, with the result that the condensation threshold is reached first of all in a very narrow region where the excitation is greatest. As the generation level is further increased, the number of excitons continues to increase on account of those regions where the condensation threshold has still not been reached. The expansion, with increasing excitation level, of the spatial region in which the EHD are concentrated has been pointed out by many workers [57, 78, 114, 117] and is also apparent from Fig. 10. At the same time, the above picture of the kinetics during the exciton condensation process accords with known results [30] on the observation of hysteresis phenomena in EHD and FE luminescence, which can hardly be explained in any other way.

It has been suggested [103] that in the steady state the electron—hole drops have the minimum stable radius R_{min}, dependent solely on temperature. In this case the exciton density must have the value n_{min} and the EHD number density the greatest value consistent with Eqs. (4.23) and (4.26), i.e.,

$$N_{max} = \frac{3\tau_0}{4\pi R_{min}^3 n_0} \left(g - \frac{n_{min}}{\tau} \right), \tag{4.36}$$

since, if the exciton density exceeded n_{min}, nuclei of new EHD would be formed, which would begin to grow and thereby reduce the exciton density; in accordance with (4.16), this would lead to a reduction in the flux of excitons to drops previously formed which had reached a size corresponding to the initial exciton density. The radius of these drops would therefore decrease. This process must go on until all drops have the same radius R_{min}. This picture does not agree, however, with the results of experiments. If this argument were valid, the EHD number density would not depend on the slope of the front of the exciting pulse nor would the EHD radius depend on excitation intensity. Also, the experimentally observed value of the EHD radius is considerably greater (especially at low temperatures) than the calculated value of R_{min}, as is apparent from a comparison of Figs. 18 and 19 with Figs. 24 and 26. In order to establish a steady state described by expressions (4.24), (4.25), and (4.36), a time enormously greater than the duration of the exciting pulses would probably be required [137].

The picture of exciton condensation described above and illustrated schematically in Fig. 29 has been discussed theoretically by Keldysh [58] from the standpoint of the theory of formation of "critical" nuclei of liquid phase in a supersaturated vapor [16, 135, 138, 139]. In his paper Keldysh derived a number of expressions describing the dependence of EHD number density on temperature, excitation intensity, and the slope of the leading edge of the exciting pulse. These expressions can account for almost all the qualitative features of our experimental results: the stronger temperature dependence of EHD number density near threshold compared with lower temperatures (Fig. 27); the different dependence of EHD number density on excitation intensity for long and short fronts and the variation of these dependences with temperature (Fig. 28); the dependence of EHD number density on the slope of the front (Fig. 25).

A detailed quantitative interpretation of the experimental results on the basis of the theoretical formulas of [58] was not attempted, since the experimental results are undoubtedly distorted by the expansion of the spatial region in which condensation occurs as a result of a decrease of temperature or an increase of excitation level (see Fig. 10). We note, however, that not only do the formulas of [58] give a qualitatively correct description of the observed phenomena; they also correctly predict the EHD number density in order of magnitude. Thus, for example, for an exciting pulse with a short front, the drop number density is given by the expression [58]

$$N \simeq \left[\frac{g}{v c_T^2 t_0}\left(\frac{2\pi\hbar^2}{M^* kT}\right)^{3/2}\right]^{3/2}\left[\frac{n_0}{v}\left(\frac{2\pi\hbar^2}{M^* kT}\right)^{3/2}\right]^2 \exp\left\{\frac{7}{2}\frac{\varepsilon_0}{kT}\left[1-2\left(\frac{2\pi}{15}\right)^{1/3}3\eta\right]\right\}, \tag{4.37}$$

where $\beta = \sigma/\varepsilon_0 n_0^{2/3}$, and the parameter η is determined by the equation

$$\frac{1}{\eta^2}-\eta = \frac{1}{10}\left(\frac{15}{2\pi}\right)^{1/3}\frac{n_0^{2/3}}{\sigma}kT \ln\left[\frac{2\Lambda}{G^2}\left(\frac{n_{0T}}{n_0}\right)^3\right]. \tag{4.38}$$

Here $G = g\tau^2/n_{0T}t_0$ and $\Lambda = 128\,(\sigma/kT)^{1/2}N_i\,(v_T\tau)^4$, where N_i is the density of impurities acting as condensation centers. Setting $t_0 = 3\,\mu\text{sec}$, $\sigma = 2\cdot 10^{-4}$ dyn/cm, $g\tau \approx 5\cdot 10^{13}$ cm^{-3} (which corresponds to an exciting power of ≈ 8 mW for light beams superposed as in Fig. 13), $\tau = 5\,\mu\text{sec}$ and $N_i \approx 10^{12}$ cm^{-3} (the values of the other parameters are cited above), an estimate using Eqs. (4.37) and (4.38) gives, for the EHD number density, $N \sim 4\cdot 10^5$ cm^{-3} at $T = 3.2°$K and $N \sim 10^7$ cm^{-3} at $T = 2.8°$K, in reasonable accord with the results shown in Fig. 25. The slope of the straight line log N vs. $1/T$ (see Fig. 25) is described rather well by expression (4.37) with $\varepsilon_0 \simeq 2$ meV, which corresponds to spectral measurements of the binding energy of the particles in the liquid phase. The temperature dependence of the EHD number density thus provides us with an explanation of why the values of the FE work function determined from spectral [26, 29, 34, 37, 86] and temperature measurements [30, 31, 46, 86] are different.

We note that the value of the coefficient of surface tension cited above may be altered if the experimental results are analyzed on the basis of the formulas of [58]; this would necessitate, however, excitation of a higher degree of uniformity.

This is an opportune moment at which to raise the following matter. When the well-known controversy concerning the nature of the luminescence line peaking at around 709.6 meV was going on, some workers considered that the quadratic dependence of the intensity of this line on generation rate which they observed was one of the most weighty arguments in favor of the biexciton nature of this emission [80, 109]. For an electron–hole drop mechanism, on the assumption that the drop number density is independent of excitation level, a cubic dependence was expected, one which was, indeed, observed in [19, 22]. It was shown above that the EHD number density is strongly dependent not only on the excitation level but also on how this level is reached. Accordingly, the differences in the experimental conditions in these papers (cw excitation in [19, 22]; pulsed excitation in [80, 109]) could account for the different dependences of luminescence intensity on excitation intensity observed by these workers. We note that a cubic dependence is observed most clearly at high temperatures. It can be seen from Fig. 26 that at high temperatures the radius of the drops is almost independent of excitation level. This would appear to indicate, therefore, that the cubic dependence is explained by the growth of the EHD number density with increasing excitation level, and not, as was suggested in [19, 40, 86], by growth of the radius of the drops.

The dependence of drop number density on generation rate accounts for the variation of the steady-state drop radius with excitation intensity at low temperatures. It can be seen from Fig. 19 that if the drop number density remains constant or changes only slightly with decreasing excitation level, then the radius of the drops must increase. If, however (as obtains for excitation by pulses with a short front), decreasing the excitation level is accompanied by a pronounced reduction of the drop number density, then the EHD radius must increase. At low temperatures and a high excitation level, when (4.31) is applicable, the radius of the drops is proportional to $(g/N)^{1/3}$ and its temperature dependence is determined solely by the temperature dependence of the drop number density. We show in Fig. 30 a plot of the ratio of the generation rate to the volume of liquid phase, g/NR^3, vs. drop radius R for three different excitation levels. For small R (low temperatures), the quantity $g/NR^3 \simeq$ const, as, indeed, follows from (4.31). Numerical estimates show that, for the generation level used by us, $g \sim 10^{20}$ $cm^{-3} \cdot sec^{-1}$ (light beams superposed as in Fig. 20), the experimentally obtained value of R^3N is almost exactly equal to the value calculated using (4.31).

At large R (high temperatures) the quantity g/NR^3 grows sharply at an almost constant value of the radius (Fig. 30), i.e., the EHD radius tends quite clearly towards a value of $\simeq 9.5\,\mu m$ (Figs. 24 and 26), which is dependent neither on EHD number density, nor on temperature, nor on the level or method of excitation. That the radius of the drops should be independent of their number density is quite understandable, since, at high temperatures, recombination in the gaseous phase dominates, and the EHD radius must be given by expression (4.32). Also, since the EHD number density for a given temperature and excitation level is greater for a short front than for a long front, it follows that the rate of recombination in the gaseous phase becomes dominant at higher temperatures in the former case; accordingly, for excitation by pulses with a short front, the drop radius saturates at higher temperatures (cf. Figs. 26a, b). The fact that the radius of the drops is nonvarying with respect to temperature and excitation intensity cannot, however, be explained on the basis of (4.32); this seems to indicate that the theoretical picture described in Section 1 is incomplete.

The experimental data thus appear to indicate that some sort of mechanism must exist which sharply curtails the growth of the drops at R ~ 10 μm, at any rate under the conditions of our experiments. The rate of growth of the drops, as can be seen from (4.10), can be curtailed as a result of a reduction in the exciton flux supplying force exerted on the excitons by the nonequilibrium phonons emitted from the drop and created within it through carrier recombination [116], or as a result of a reduction of the exciton diffusion coefficient with increasing exciton density on account of exciton–exciton collisions.

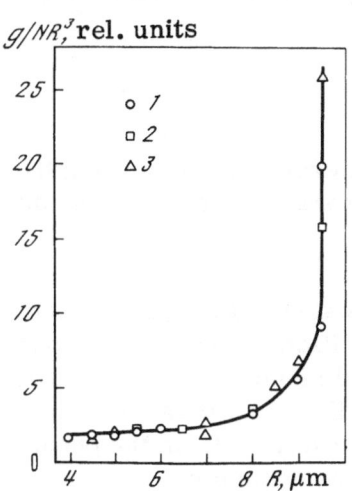

Fig. 30. The quantity g/NR^3 plotted vs. EHD radius for three different excitation intensities. (1) 8 mW; (2) 3.8 mW; (3) 2.4 mW.

We allow for the effect of exciton−exciton collisions by writing the exciton diffusion coefficient in the form $D_{ex} = {}^4/_3[l_el_0/(l_e + l_0)]v_T$, where l_0 is the exciton mean free path ignoring exciton−exciton collisions, and $l_e = {}^1/\sigma_en(r)$ is the exciton mean free path determined by collisions of excitons with one another (σ_e is the cross section for exciton−exciton scattering; $n(r)$ is the exciton density, a function of the distance r to the center of the drop). The factor 4 in the expression for D_{ex} arises because v_T, the thermal velocity, determines the pressure of the exciton gas and not the mean velocity. The exciton diffusion coefficient thus decreases with increasing n. Inserting the above expression for D_{ex} into (4.12) and remembering that we require to find the exciton density in a small region near the drop, we obtain, instead of (4.14), the expression

$$\frac{d}{dr}\left[r^2\frac{l_0}{1 + l_0\sigma_en(r)}\frac{dn(r)}{dr}\right] = 0. \tag{4.39}$$

Solving this equation subject to the boundary conditions (4.11) and $n(r \to \infty) = n$ and the steady-state condition (4.21), we obtain, instead of (4.23), the following expression for the mean exciton density as a function of the radius of the drop:

$$n = e^{R^2/R_0^2}\left[n_T(R) + \frac{n_0R_0^2}{3D\tau_0}\left(1 + \frac{D}{v_TR}\frac{R^2}{R_0^2} - e^{-R^2/R_0^2}\right)\right], \tag{4.40}$$

where D is the exciton diffusion coefficient ignoring exciton−exciton collisions, and

$$R_0^2 = 4v_T\tau_0/n_0\sigma_e. \tag{4.41}$$

A calculation of σ_e allowing for the real band structure of Ge, the anisotropy of the effective masses, and the presence of "heavy" and "light" excitons, was not attempted. However, in the case of isotropic and nondegenerate bands, the contribution of exciton−exciton collisions must be negligibly small as momentum conservation implies that such collisions do not lead to a reduction of the exciton flux. In the case of the real band structure, the effect of exciton−exciton collisions is unlikely to be large enough to explain the experimentally observed limiting value of the EHD radius, which would require a value of σ_e of the order of the geometrical cross section of the exciton.

We briefly consider the entrainment of excitons in the phonon wind from the drops [116]. We assume that each act of recombination in the drop releases an energy of the order of the bandgap E_g and that recombination goes mainly through nonradiative channels. A flux of energy given by

$$w = \frac{4\pi}{3}R^3\frac{n_0}{\tau_0}E_g. \tag{4.42}$$

then emanates from the drop. These phonons collide with the excitons surrounding the drop and are scattered by them, as a result of which excitons are carried away from the drop.[†] The

---

† Carrier recombination in the drops produces mainly short-wavelength acoustic phonons, which can travel quite a distance from the drop during their lifetime (optical phonons decay into acoustic very rapidly). It can be shown from energy and momentum conservation that the absorption of such phonons by excitons is negligible small. Also, it can be shown using the conservation laws that the momentum transferred to the excitons from the phonons as a result of collisions is of the order of the exciton thermal momentum p_T.

force acting on an exciton on account of the "phonon wind" is given by the expression

$$\mathbf{f} = \frac{1}{3} \frac{n_0}{\tau_0} \frac{E_g \sigma_p p_T}{\hbar \omega} \frac{R^3}{r^3} \mathbf{r}, \tag{4.43}$$

where $\hbar \bar{\omega}$ is the mean phonon energy, σ_p is the cross section for exciton − phonon scattering, and $p_T = (2m^*_{ex} kT)^{1/2}$ is the exciton thermal momentum. The density of the exciton flux, when allowance is made for the diffusion of the excitons and their directional drift under the action of the force $\mathbf{f}$, is then given by

$$S(r) = -D \nabla n(r) - \frac{D}{kT} \mathbf{f}(r) n(r), \tag{4.44}$$

where D/kT is the exciton mobility. In order to find the mean exciton density as a function of EHD radius under steady-state conditions, we are required to solve the equation div S = 0 subject to the same boundary conditions as for (4.39). The solution is expression (4.40) but with R_0 given by

$$R_0^2 = 3 \sqrt{\pi} \frac{v_T \tau_0}{n_0 \sigma_p} \frac{\hbar \bar{\omega}}{E_g}. \tag{4.45}$$

The cross section σ_p is calculated in [116]. Unfortunately, the absence of information on the frequency distribution of the phonons emitted from the drops makes reliable estimation of R_0 difficult. The estimates made in [116] give for R_0 a value of ~100 μm, which is greater than the experimental value. Despite the quantitative discrepancy, the removal of excitons by phonons nonetheless gives a qualitatively correct description of the behavior of the EHD radius. For $R \ll R_0$ expression (4.40), as is to be expected, goes over into (4.23). As $R \rightarrow R_0$, however, the exciton density, as can be seen from (4.40) and Fig. 31, increases sharply [~exp (R^2/R_0^2)]; in other words, the growth of the radius of the drops with increasing excitation level is sharply curtailed, in qualitative agreement with experiment. Also, the temperature dependence of the EHD radius proves to be very weak, since, by (4.41) and (4.45), $R_0 \sim T^{1/4}$.

A consequence of the weak dependence of the radius of the drops on exciton density when $R \sim R_0$ is that the sizes of the drops tend to equalize out at large R, i.e., the drop size distribution must become more uniform with increasing drop radius. This may be why the calculated points are in better agreement with the light-scattering curves at high temperatures (see Fig. 13).

The phonon wind (or exciton−exciton collisions) makes it much more difficult for the drops to grow via condensation of excitons from the gaseous phase. It should be noted, how-

Fig. 31. Graphs of n vs. R calculated using Eq. (4.40) for three different temperatures with $R_0 \simeq 15$ μm at T = 2°K and calculated using formula (4.26) for three values of EHD number density N. The numerical values of the parameters are the same as used in the calculation of the curves shown in Fig. 18.

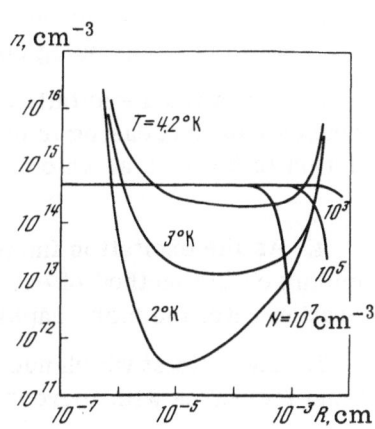

ever, that this does not preclude the formation of drops of size $R \gg R_0$ [116]; for example, as a result of carrier generation directly within a drop, or if a carrier density $n \gtrsim n_0$ is produced suddenly in some part of the sample by a sufficiently intense exciting pulse. Neither are our conclusions at variance with the results of [51, 118, 119], in which EHD with a radius of hundreds of microns were observed. The fact is that these large drops were produced in uniaxially strained samples and had much smaller carrier densities n_0 and, as a result [118], much greater lifetimes ($\sim 5 \cdot 10^{-4}$ sec). And it is readily seen from (4.41) and (4.45) that reducing n_0 by an order and increasing τ_0 by an order has the effect of increasing R_0 by an order.

Near the condensation threshold, when the quantity n_{min}/τ becomes comparable with the generation rate, the EHD radius (as is apparent from Figs. 18, 19, 31) must decrease with increasing temperature and reach the value R_{min} at the threshold temperature. In our experiments, as T approached the threshold temperature, the decrease in the radius of the drops was not more than 20% of its greatest value. This is probably because the threshold temperature, in actuality, can never be reached: As T approaches threshold temperature, the "critical" radius increases strongly and the probability of nuclei being formed is decreased. It may be that this is why the threshold temperature in impurity samples, where condensation centers are present, is always higher than in pure samples for a given generation rate [22].

The experimental results can thus be said to be in reasonable qualitative agreement with the theoretical picture described above. A complete quantitative description of the observed phenomena is probably too much to expect, since, firstly, under the conditions of the experiments, the nonequilibrium carriers are insufficiently uniformly distributed over the volume of the sample, and, secondly, a number of the parameters are in need of direct experimental determination.

MAIN RESULTS AND CONCLUSIONS

1. The scattering of light by EHD in germanium is studied using a high-sensitivity setup permitting automatic recording of the scattering indicatrix. Light absorption by nonequilibrium carriers and the spectrum of the Ge recombination luminescence are recorded simultaneously with the scattering measurements. High sensitivity was achieved by using a laser amplifier to amplify the scattered light, laser volume photoexcitation of the sample, and a relatively high-power laser source of probing light.

2. The parameters describing the basic properties of an exciton gas – EHD system are measured: carrier density of the electron – hole liquid; size and number density of the EHD; mean density of nonequilibrium carriers in the part of the crystal occupied by the system. Under our experimental conditions, the radius of the droplets of condensed phase varied from 3 to 12 μm in the temperature range 1.9 to 4.2°K while their number density varied from 10^8 to $3 \cdot 10^3$ cm^{-3} depending on temperature, excitation intensity, and method of excitation.

3. For a fixed excitation level, the size of the EHD increases and their number density decreases with increasing temperature. The decrease of number density is explained by the reduction in the rate at which viable liquid-phase nuclei are formed with increasing temperature.

4. As the excitation intensity is varied, the radius of the drops can increase or decrease depending on the method of excitation; the drop number density decreases with decreasing generation rate, the more rapidly the higher the temperature.

5. The rate at which nuclei of liquid phase are formed depends on the method of excitation and increases with increasing supersaturation of the exciton "vapor."

6. At low temperatures the steady-state size of the EHD is determined by their number density, i.e., by the nucleation conditions; at high temperatures the size of the EHD is almost independent of their number density and temperature and is weakly dependent on excitation level.

7. The data obtained show that the process of nucleus formation can be controlled, i.e., it is possible to "introduce" into a crystal a definite number of EHD of prescribed size by varying the method of excitation, the excitation intensity, and the temperature of the crystal.

8. The results obtained lead to the conclusion that the kinetics of exciton condensation, and thus also the dependence of the volume of liquid phase on excitation level and temperature, are determined mainly by the process responsible for the formation of nuclei. The discrepancy between the results of spectral and temperature measurements of the carrier binding energy in the liquid phase can be understood on this basis.

9. The formation of nuclei of liquid phase and the growth of EHD are studied theoretically on the basis of a model allowing for the surface tension of the electron−hole liquid and for the diffusion of excitons toward the surface of the EHD.

10. The coefficient of surface tension of the electron−hole liquid is measured: $\sigma \simeq 1.8 \cdot 10^{-4}$ dyn/cm.

In conclusion I should like to express my warmest thanks to my supervisors, L. V. Keldysh and V. S. Bagaev, for inestimable assistance, innumerable discussions, and their constant scientific guidance throughout the work. It is a pleasure also to thank B. M. Vul, director of the laboratory, for his constant interest, N. A. Penin for his assistance in the first stages of the work and useful advice, and N. V. Zamkovets and V. A. Tsvetkov, whose contribution to the execution of this work it is difficult to overestimate.

LITERATURE CITED

1. L. V. Keldysh, Proc. Ninth International Conf. on Semiconductor Physics [in Russian], Izd. Nauka, Leningrad (1968), p. 1387.
2. L. V. Keldysh, Usp. Fiz. Nauk, 100, 514 (1970).
3. L. V. Keldysh, in: Excitons in Semiconductors [in Russian], Izd. Nauka, Moscow (1971), p. 5.
4. V. M. Asin and A. A. Rogachev, Pis'ma Zh. Éksp. Teor. Fiz., 9:415 (1969).